Molar Volumes and Jone Dole Coefficients of DIHYDROPYRIMIDINES and its Derivatives

THE AUTHORS

Dr. Manoj Ramesh Gaware has completed M.Sc. in Organic Chemistry from Savitribai Phule Pune University in 2003 and has been awarded Ph.D. by Savitribai Phule Pune University in 2017. Recently he is working as Associate Professor in Department of Chemistry, Maratha Vidya Prasarak Samaj's G.M.D Arts, B. W. Commerce and Science College, Sinnar. He is Recognised guide by Savitribai Phule Pune University, Pune. He has total 17 years teaching experience. He has published 37 research papers in reputed International Journals indexed in Scopus and Web of Science. He has completed 01 minor research project funded by UGC, New Delhi. He has presented 08 papers at different National and International seminars, conferences and workshops. He has published one International book. He has published a patent on "Medicinal Plant leaf extracts that are used to treat plant diseases were put up against each other in a physicochemical test". He has attended various national and international seminars, workshops and conferences. He has been awarded by 01 state and 05 National award in the education field. He has been invited as resource person in different colleges. As a part of community service he has started soil and water testing laboratory in the college.

Miss. Ankita Mukund Rayate has completed M.Sc. in Organic Chemistry from Savitribai Phule Pune University in 2018 and has registered for Ph. D. in Savitribai Phule Pune University. Recently she is working as Assistant Professor in Department of Chemistry, Maratha Vidya Prasarak Samaj's G.M.D Arts, B. W. Commerce and Science College, Sinnar. She has total 03 years teaching experience. She has presented papers at National and International seminars, conferences and workshops and has published a paper in International journal. She has published a patent on "Medicinal Plant leaf extracts that are used to treat plant diseases were put up against each other in a physicochemical test". She has attended various national and international seminars, workshops and conferences. As a part of community service she has contributed to different activities organized by college.

Molar Volumes and Jone Dole Coefficients of DIHYDROPYRIMIDINES and its Derivatives

– Authors–

Manoj Ramesh Gaware

Ankita Mukund Rayate

2023

Scholars World®

A Division of

Astral International Pvt. Ltd.

New Delhi – 110 002

ISBN 9789359191584

Published by : **Scholars World**
A Division of
Astral International Pvt. Ltd.
– ISO 9001:2015 Certified Company –
4736/23, Ansari Road, Darya Ganj
New Delhi-110 002
Ph. 011-43549197, 8130496929
e-mail: info@astralint.com
Website: www.astralint.com

Preface

We the authors of these books have shown scope for reinvigorating their presentation of chemistry with new ideas by studying the solution properties of organic compounds. We wanted to write a book whose structure grows from the development of ideas rather than being dictated by the sequential presentation of facts. We believe that students benefit most of all from a book which leads from familiar concepts to unfamiliar ones, not just encouraging them to know but to understand and to understand why. We also knew that if we did this we could,from the start, relate the chemistry we were talking about to the two most important sorts of chemistry that exist—the chemistry that is known as life, and the chemistry as practised by chemists solving real problems in laboratories.

We aimed at an approach which would make sense to and appeal to today's students. But all of this meant taking the axe to the roots of some long-standing textbook traditions. The best way to find out how something works is to take it apart and put it back together again. Students can soon make sense of chemistry which may be unfamiliar in detail by relating it to what they know and understand.

The authors have carefully noted and none was over looked while writing. We would also like to acknowledge the support and guidance of the publication team and again to recognize the seminal contribution of the man who first nurtured the vision could be taught. The time spent on the preparation of this book was made available only with the forbearance of our families, friends and research groups, and we thank all of them for their patience and understanding.

Manoj Ramesh Gaware

Ankita Mukund Rayate

Contents

Chapter 1

General Review

1.1 Introduction

Heterocyclic compounds are important in organic chemistry as they contribute about one third of research publications. Carbocyclic compounds are carbon atoms in the ring form. Heterocyclic compounds contain at least one atom other than carbon [1-8]. Heterocyclic compounds contain heteroatoms like Oxygen, Nitrogen and Sulphur They are either aliphatic or aromatic in nature. Aliphatic heterocyclic compounds are cyclic like amines, thioethers, ethers, alcohols *etc.* The properties of such compounds are generally affected by different factors such as ring stain. The aromatic heterocyclic compounds are similar in properties of benzene and follow (4n +2) π e^- rule also known as Hückel rule for aromatic compounds. Heterocyclic saturated compounds namely piperidine, tetrahydrofuran typically acts like amines and ethers respectively. Hence heterocyclic chemistry studies are mainly concentrated on unsaturated analogs containing unstrained rings namely furan, pyrrole, pyridine and thiophene. Indole, quinolone, benzofuran and benzothiophene are compounds containing benzene ring fused with pyrrole, pyridine, furan and thiophene. Due to presence of angle strain in three and four membered heterocyclic compounds, they are highly reactive and unstable. Due to aromaticity, five membered unsaturated rings containing one heteroatom are more stable than five membered heterocyclic compounds containing one hetero atom such as pyrrolidine, tetrahydrofuran and tetrahydrofuran.

Heterocyclic compounds have wide application and are present in nature such as vitamins, drugs, biological vital compounds possessing certain medicinal activities [9] including effective action against virus, bacteria, fungi, insect and cancer. They also possess anti HIV, anti-inflammatory, herbicidal properties also.

Some heterocyclic compounds has also found application in material science having brightening agent, dyestuff *etc.* Heterocyclic compounds play an important role in pharmaceuticals as well as agrochemicals compounds.

Various synthesized compounds used in industrial sector are also heterocyclic in nature. Such synthetic heterocyclic compounds have wide application in rocket propellants and photography.

Pyrimidines are important heterocyclic moiety in many organic compounds and contributed due importance in pharmacological applications [10], biological uses, [11-20], herbicidal effects [21], pesticides impact [22] synthetical applications [23], and polymeric and material sciences [24-27]. Pyrimidines and fused pyrimidines is an essential part of DNA and RNA plays a vital role in several biological processes. The thiouracil derivatives can be utilized as a component and adhesive for noble metals in many fields such as medical treatment, electronic materials, precision instruments and jewellery where resins are bonded to metals and are particularly useful in the dental field. Hence high substantial attention is given towards synthesis, characterization and densitometric and viscometric studies followed by theoretical or computational study of pyrimidine and its analogs such as pyrimidine carbonitriles, dihydropyrimidine-2(1H)-ones and pyrimidine carboxylates.

1.2 Electromagnetic Radiation and Spectroscopy

Organic chemist use spectroscopy as an essential and important tool for structure determination. Spectroscopy may be defined as study of the quantized of electromagnetic radiation with matter. Electromagnetic radiation is produced due to oscillation of electric charge and magnetic field residing on atom. There are various forms of electromagnetic radiation such as light (visible), infrared, ultraviolet, X-rays, microwaves, radiowaves, cosmic rays, *etc.*

1.2.1 Ultraviolet (UV) and Visible Spectroscopy [28, 29]

Ultraviolet and Visible spectroscopy deals with recording of absorption of radiations in ultraviolet and visible region of the electromagnetic spectrum. The ultraviolet region extends from 10-400 nm while visible region extends from 400-800 nm. The absorption of electromagnetic radiation in UV and visible regions induces excitation of electron from lower to higher molecular orbitals. Hence it is also termed as electronic spectroscopy. When continuous radiation pass through transparent medium, a portion of this radiation may be absorbed which results in absorption of energy. Ultra-violet spectroscopy is such type of spectroscopy which involves absorption of electromagnetic radiation in the UV-visible region of electromagnetic spectrum region (200-800 nm). This spectroscopy is also called as electronic spectroscopy and involves electronic transitions. Due to absorption of electromagnetic radiation, electron is promoted from occupied orbital to an unoccupied orbital of greater potential energy that is excitation of valance electron from lower electronic energy level to higher electronic energy level (Σ, π and n to Σ^*, π^*). The energy differences between electronic levels in most of the molecules are in between 125-650 KJ mol^{-1}. This spectroscopy has main application for detection of nature of conjugated multiple bonds and aromatic rings. Polar groups are stabilized by polar solvents by hydrogen bonding or dipole-dipole interactions also known as London forces. Chromophore is defined as group of atoms which absorbs characteristic energy of transition and wavelength of energy rather than electrons themselves. In ground state polarity of chromophore is high then in excited state therefore stabilization of polar solvent is more in ground state as compared to excited state. Due to change in spectral band position in the absorption, reflectance, transmittance and emission, the spectrum of molecule is shifted to lower or shorter wavelength called as hypsochromic shift which intensify the polarity induced in solvents. Chromophore bearing polar group in excited state than there is change in spectral band position in the absorption, reflectance, transmittance and emission and the spectrum of molecule is shifted to higher or longer wavelength called bathochromic shift which is observed and there is increase in polarity of solvent which usually shifts n to π^* and Σ^* bands to lower wavelength and π to π^* bands to higher wavelength.

Application of UV Spectroscopy

(i) To Determine Effect of Conjugation and Nature

The extent of conjugation in double bonded system increases bathochromic shift. Conjugation of two chromophore not only results in bathochromic shift but increases the intensity of absorption. These two effects are used to study interpretation of ultraviolet spectra of organic molecules. The exact position and intensity of absorption band of the conjugated system can be correlated with the extent of conjugation in the system.

(ii) Electron Releasing and Electron Withdrawing Effect

Substituents have different effects on the position of absorption maxima depending upon whether they are electron donating or electron withdrawing. Any groups shift the primary absorption band to longer wavelength independent to its influence on electron distribution except aromatic compounds. Electron withdrawing group have no effect on the secondary absorption and is capable of acting as chromophore. However electron donating group increases wavelength as well as intensity of secondary absorption band.

(iii) Colour in Compounds

The colour observed is due to light emitted from source corresponding to the wavelength. When light falls on object, light of particular wavelength is absorbed while remaining of it is reflected. Hence some compounds serves as relationship between absorption spectra and observed colour. Extension conjugation shifts spectra to longer wavelength that they absorb visible light appear to be coloured. *e.g.* β-Carotene is orange in colour and absorbs at 452 nm, Cyanidin is blue in colour and absorbs at 545 nm.

(iv) Geometrical Isomers

Due to absence of planarity of isomers having steric hindrance gives lost of conjugation and results in shifting of spectra to lower or shorter wavelength. Therefore cis isomer have high steric hindrance due to presence of close vicinity of groups absorbing at lower wavelength than trans isomer.

1.2.2 IR Spectroscopy (Infrared Spectroscopy) [28, 29]

Unlike other energy absorption, molecules are excited to higher energy state, when they absorb infrared radiation. Infrared spectroscopy deals with recording of the absorption of radiations in infrared region of electromagnetic radiation. The position of Infrared absorption is expressed as wavelength in micron (μ) nor wavenumber (á) cm^{-1} which is directly proportional to radient energy. When molecule is subjected to IR radiation, results in increase in amplitude of the vibrational motion of the bonds of molecule which possess permanent dipole moment causes excitation of molecule from lower vibrational level to higher vibrational level containing number of closely packed spaced rotational levels resulting in vibrational and rotational energy changes. Change in dipole moment increases, intensity of absorption also increases. Therefore infrared spectroscopy is also known as vibrational- rotational spectroscopy and the spectrum is known as vibrational-rotational spectrum. Due to numerous adsorption bands, more structural information can be withdrawn about the molecule. Absorption bands in IR spectrum are correlated different bonds and functional groups. Absorption band below 1500 cm^{-1} is called finger print region which helps to differentiate two compounds except enantiomers in which absorption pattern in IR spectrum in this region is not exactly same. The vibration in each functional group is called fundamental absorptions and absorbs particular frequency and strength. Hence IR spectroscopy finds an application in distinguishing and to identify organic compounds. In general 4000-400 cm^{-1} region is very useful in prediction of structure of organic compound.

The modes of vibrational motion in molecule that are infrared active responsible to give rise to absorptions are "stretching" and "bending". In general asymmetric stretching vibrations occur at higher frequencies than symmetric stretching vibrations. At the same time, stretching vibrations occurs at higher frequency than bending vibrations. Other terms such as scissoring, rocking, wagging and twisting are commonly used to describe the origins of infrared bands.

The plot of absorption intensity against wavenumber or sometimes wavelength is referred as infrared spectrum. The absorption bands are referred as strong (s), medium (m), weak (w), broad or sharp Each interatomic bond vibrates in several different modes like stretching or bending. Each individual bond absorbs more than one IR frequency. Stretching adsorption mode produces strong peak than bending but weaker bending absorption can be useful for differentiating similar type of bonds. Molecular vibrations depend on different parameters such as interatomic distances, bond length, bond angle, bond strength *etc.* Hence these vibrational frequencies provide molecular fingerprint useful for identification in pure as well as mixture state. Therefore IR spectrum is very useful for organic chemist for structure determination.

Applications of Infrared (IR) Spectroscopy

(a) Functional Group Determination

The basic information about the functional groups or type of vibrations abstracted can be interpretate easily from IR spectrum. From the spectral data, functional group and other informations can be evaluated for identifying organic compounds.

(b) Conjugation Effect

Conjugation of a C=C bond with carbonyl or another double bond provides the multiple bond with more single bond character through resonance or mesomeric effect lowers value of force constant "K" and ultimately lowering of vibration. *e.g.* Vinyl double bond absorbs at 1630 cm^{-1}.

(c) Hydrogen Bond

In alcohols and phenol broad O-H stretching vibration is obtained due to intermolecular hydrogen bonding in the range from 3400-3300 cm^{-1}.If alcohol is diluted with carbon tetrachloride, a sharp "free" (non-hydrogen bonded) O-H band appears at about 3600 cm^{-1} to the left of broad band. When the solution is further diluted, the broad intermolecular hydrogen- bonded band is reduced leaving behind major band of free O-H stretching absorption. Therefore intermolecular hydrogen bonding weakens the O-H bond thereby shifting band to lower frequency. These distinctions provide useful information.

Intramolecular hydrogen bonding usually shifts O-H to lower frequency. The intramolecular hydrogen bonding doesn't change its position significantly even at high dilution because the internal bonding is not altered by change in concentration.

(d) Progression in Reaction

Study of progress or completion of reaction can be interpreted from IR spectroscopy by withdrawing a small portion of reaction mixture over variable span of time. *E.g.* Conversion of secondary alcohol to ketone can be studied from observing adsorption bands in IR spectrum. The disappearance of absorption band at 3570 cm^{-1} due to O-H and appearance of absorption peak at 1720 cm^{-1} due to (-C=O) group confirms completion of reaction.

Different Orientations in Aromatic Compounds

Out of plane bending vibrations provide more information than in-plane bending vibrations in monosubstituted aromatic system. Ortho substituted compounds show one strong absorption band at 750 cm^{-1} while meta substituted compounds have absorption bands at 690, 780 and 880 cm^{-1} and para substituted aromatic compounds show absorption band at 800-850 cm^{-1}.

Per cent Purity Analysis of Compound

Impurity free organic compounds gives samples give clear IR spectrum and impure compounds give IR with many bands which is uninterpreted. *e.g.* ketonic impurity in hydrocarbon sample can be analysis by observing band at around 1720 cm^{-1}. Also per cent composition of particular component can be deduced from intensity of absorption peak. Therefore IR spectroscopy is strong tool for determining percentage purity of sample by interpretating with standard one.

1.2.3 NMR Spectroscopy (Nuclear Magnetic Resonance Spectroscopy) [28, 29]

Nuclear magnetic resonance also known as NMR is a more important spectroscopic technique than infrared spectroscopy. Many nuclei can be studied by NMR technique but hydrogen and carbon are most commonly used. NMR spectrum gives information about the number of magnetically distinct atoms to be studied. In case of hydrogen nuclei, the number of each of the distinct type of hydrogen nuclei as well as information regarding nature of the immediate environment can be known. NMR along with IR helps to determine the complete structure of unknown organic compound.

(a) Nuclear Spin Rate and Chemical Shift

Many atomic nuclei have unique property of spinning and possess quantized spin angular momentum and magnetic moment. In absence of an applied magnetic field, all the spin states of given nuclei are equivalent in energy (degenerate state) and almost equally populated with the same number of atoms having allowed spin. In presence of applied magnetic field, all protons have their magnetic moment either aligned with the field or opposite to it. If the magnet oppose (non-aligned) it repeals the current orientation. Hence due to external magnetic field, the degenerate spin state splits into two unequal energy state.

The nuclear magnetic resonance phenomenon occurs when the nuclei is aligned with an applied field are induced to absorb energy and change its spin orientation with respect to the applied field and the nucleus begin to process about its own axis of spin with angular momentum "ω" called "Larmor frequency" which is proportional to strength of applied field. Since the nucleus bear charge, the precession generates an oscillating electric field of same frequency. If the radiofrequency supplied matches with frequency of oscillating nucleus, there is coupling of these two field and energy is transferred to the nucleus causing spin change This condition is called "resonance".

It is very difficult to measure exact frequency than precision instead a reference compound is placed in the solution of the substance to be measured and the frequency difference is measured. The standard reference substance used is tetramethyl silane [$(CH_3)_4Si$] also known as TMS. The shift from TMS for given proton depends on strength of applied magnetic field. The chemical shift (δ) expresses the amount by which proton resonance is shifted from TMS in parts per million (ppm) of the spectrometers basic operating system. On the scale the resonance of all protons of TMS comes exactly at 0 ppm.

Chemical shift depends on various factors such as inductive effect, anisotropic effect, hydrogen bonding and van der Waals forces.

In NMR spectrum, protons which appears in right side are highly shielded and termed as upfield region or high field region and that appearing to the left side are highly deshielded is termed as down field region or low field region.

All the protons present in chemically identical environment within a molecule are chemically equivalent and exhibit same chemical shift. On the other hand molecules possessing different set of protons give rise to different absorption peaks and the set of protons are chemically non-equivalent.

(b) Coupling Constant (J)

The splitting pattern and the ratio of intensities in multiple component doesn't explain the quantitative amount of splitting pattern of peak. The distance between the peaks in the simple multiplet is called as coupling constant (J) and is always expressed in Hertz (Hz). It provides interrelationship between neighbouring nuclei and chemical environment in which particular proton is present. It is used to know the information regarding NMR active nuclei and geometrical isomerism (cis-trans).

1.4 Solution Chemistry

Many interactions along with reactions performed in research, industry as well as environment takes place in solution state. Solutes are generally dispersed in solids, liquids and gases. Over a period of time, the solvent effect has been ignored frequently and desperately. From last few decades the chemists were concerned mainly with interactions of water only.

In solid state, the molecules are arranged in perfect manner while in gaseous state molecules are dispersed and far away from each other while in liquids the state is an intermediate between these two. The kinetic theory of gases explains about kinetic motion and drastic collisions between the atoms and also with the container within short distances. to atoms. Liquids possess very small space between the molecules or atoms hence random movement is somewhat restricted and also the energy is dissipated by perfectly elastic molecular collision.

Collisions of continuously moving molecules dissipate energy along with energy received from neighbouring molecules. The average K.E (kinetic energy) of molecule is directly proportional to the absolute temperature. Liquid moves slowly and are held together by intermolecular forces in particular volume. But due to strong molecular forces, the molecules are held tightly at any position. Hence liquid possess proper shape, but due to elastic property of liquid molecule, they can simply proceed resulting in shape of container. Density of gases is less as compared to liquids due to close vicinity of molecules. The temperature and pressure change effects very negligibly on liquid volume due to compactness between liquid molecules

Liquid motion may be streamline or laminar way [35]. If exact path is followed by every particle of liquid passing through a particular point along a smooth path gives streamline or laminar flow. Contradictory, the flow may become turbulent or irregular causing change in velocity. This non-uniform movement of liquid is called as eddy current.

Densitometric and Viscometric parameters gives arrangement of molecules within liquid and helps to derive thermodynamical properties which are essential for interpreting extent of strength and type of molecular interactions within the solutions.

1.5 Density and Viscosity

Density (ρ)

"Density" means mass (m) to volume (v) ratio and expressed in g/cm^3 in CGS system or kg/m^3 in MKS system. Due to thermal expansion, the volume of matter increases with increase in temperature decreasing the density of the matter except water which possess anomalous behaviour *i.e.* density increases with increase in temperature from 0-4°C. Density is dependent on pressure exerted by gaseous substances. Thermodynamic parameters such as apparent molar volumes evaluated from density measurement helps to predict solute-solvent interactions and also provide information about the extent of strength and type of the molecular interactions.

Viscosity (s)

Another term "viscosity" is direct method for determining quality and changes in substance under observation [36]. It is related to resistance offered by material to flow. This is due to the fact of intermolecular attractions of the liquid and depends on the type as well as molecular weight of the liquid.

Viscometric studies are of great importance in agro based industries in which exact idea of viscosity is required for performing different methods [37]. The viscous nature of the liquid is due to shearing effect that is movement of liquid layers over each other. Hence resistance to flow is observed while sliding over one another [38-42]. Higher the resistance, more viscousness is offered by liquid to flow and ultimately more is the viscosity of the liquid. Elevation or rise in temperature raises randomness in movement of liquid molecule which ultimately reduces difference between velocities of adjacent layer decreasing viscosity of liquids.

Viscometric studies reveal an idea regarding solute-solute interactions in aqueous as well as in non-aqueous solutions.

Viscosity of liquid varies with composition along with temperature and concentration which is main objective about our study. Therefore this is the cause to study how viscosity of ternary system is related to concentration and temperature.

Different viscometric methodologies known, there are two distinct ways namely kinematic viscosity and Dynamic or absolute viscosity. Kinematic viscosity measures resistance offered by liquid to flow under gravity while dynamic or absolute viscosity measures resistance to flow under external controlled force by capillary or body moving through liquid. Unit of kinetic viscosity is centistokes (cSt) while unit of dynamic viscosity is centipoise (cP) for comparable result, dynamic viscosity is transformed into kinetic viscosity by fractionating with specific gravity.

1.6 Present-Day work

According to our knowledge, densitometric, viscometric and Hartree-Fork/ Density Functional Theory frequency calculations studies of pyrimidines and its derivatives such pyrimidine carbonitriles, dihydropyrimidin-2(1H)-ones, pyrimidine carboxylates have not been reported so far. Hence the present investigation consists the study of density and viscosity in binary liquid mixture of dimethyl sulphoxide and water along with study structure forming or structure disintegrating behaviour of dihydropyrimidine carbonitriles, dihydropyrimidin-2(1H)-ones and dihydropyrimidine carboxylates and its analogs by determining the apparent molar volume from densitometric and "'A" and "B" coefficients from viscometric data followed by theoretical or computational study including geometry optimizations such as bond length, bond angle, dihedral angle, Mulliken's charge, FTIR vibrational spectrum and 1H-NMR spectrum of these molecules completely for identification of various normal modes with high wave number accuracy using Gaussian 09 software.

1.7 References

1. J. H. Fletcher, O. C. Dermer and R. B. Fox B; *"Nomenclature of Organic Compounds, Principles and Practice", American Chemical Society Washington, D. C., Adv. Chem. Ser.,* **1966**.
2. J. A. Joule and G. F. Smith, *"Heterocyclic Chemistry", Van Nostrand Reinhold Co. 2nd Ed, London,* **1978**.
3. A. R. Katritzky; *"Physical Methods in Heterocyclic Chemistry"*, Vol.1-5, *Academic Press, New York,* **1963-1973.**
4. A. R. Katritzky, *"Comprehensive Heterocyclic Chemistry"*, Vol.1-8.
5. G. R. Newkome and W. W. Pandler; *"Contemporary Heterocyclic Chemistry", John Wiley, New York,* **1982**.
6. E. Campaigne, *Adrien Albert and the Rationalization of Heterocyclic Chemistry", J Chemical Education,* Volume 6, **1986**, 860.
7. T. L. Gilchrist, *"Heterocyclic Chemistry", Pitman London,* **1985**.
8. A. R. Katritzky, *"Handbook of Heterocyclic Chemistry", Pergamon Press, New York,* **1985**.

9. U. Sehlstedt, P. Aich, J. Bergman, H. Vallberg, B. Norden and A. Graslund, *Journal of Molecular Biology.* 278, **1998**, 31.

10. C. O. Kappe *Tetrahedron.* 49, **1993**, 6937 and references cited therein.

11. (a) Patil A D, N. V. Kumar, W. C. Kokke, M. F. Bean,A. J. Freyer, De Brosse, C Mai, A. Truneh, D. J. Faulkner, B. Carte, A. L. Breen, R. P. Hertzberg P, R. K. Johnson, J. W. Westley and B. C. Potts, *J. Org. chem.* 60, **1995**, 1182.

    (b) B. B. Snider, C. J Patil, A D and A. Freyer *Tetrahedron. Lett.* 37, **1996**, 6977.

12. J. Clark, M. S. Shahhet, D. Korakas and G. J. Varvounis. *Het. Chem.* 30, **1993**, 1065-1072.

13. K. Ogowva, I. Yamawaki, Y. I. Matsusita, N. Nomura, P. F. Kador and J. H. Kinoshita *Eur. J. Med. Chem.* 28, **1993**, 769-781.

14. B. Tozkoparan, M. Ertan, P. Kelicen and R. Demirdar, *Farmaco* 54, **1999**, 588-593.

15. M. Santagati, M. Modica, A Santagati, F. Russo and S. Spampinato, *Pharmazie.* 51, **1996**, 7-11.

16. V. K. Ahluwalia, M. Chopra and R. A. Chandra, *J. Chem. Rs.* 5, **2000**, 162-163.

17. M. Van Laar, E. Volkerts and M. Verbaten, *Psychopharmacology.* 154, **2001**, 189-197.

18. K. Danel, E. B. Pedersen and C. J. Nielsen. *Med. Chem.* 41, **1998**, 191-198

19. O. A. Fathalla, S. M. Awad and M. S. Mohamed *Arch. Pharm. Res.* 28, **2005**, 205–1212.

20. O. A. Fathalla, A. Zaghary, H. H. Radwan, S. M. Awad and M. S. Mohamed. *Arch. Phar. Res* 25, **2002**, 258-269.

21. Y. Ding, J. Girardet, K. L. Smith, G. J. Prigaro, J. Z. Wu and N. Yao *Bioorg. Chem.* 34, **2006**, 26–38.

22. T. Azza and M. A. Sahar, *Molecules.* 17, **2012,** 9868-9886.

23. M. Kimura and M. Aizawa, *USP 15005795497161A* 5, **1998,** 795,497.

24. P. Moreau, F. Anizon, M. Sancelme, M. Prudhomme, C. Bailly, C. Carrasco, M. Ollier, D. Severe, J. Riou, D. Fabbro, T. Meyer, A. Aubertin, A. *J. Med. Chem.* 41, **1998,** 1631.

25. V. Danilenko, A. Simonov, S. Lakatosh, M. Kubbutat, F. Totzke, C. Schachtele, S. Elizarov, O. Bekker, S. Printsevkaya, Y. Luzikov, M. Reznikova, A. Shtil, S. Preobrazhenskaya, *J. Med. Chem.* 51, **2008,** 7731.

26. A. Kozikowski, I. Gaisina, P. Petukhov, J. Sridhar, L. King, S. Blond, T. Duka, M. Rusnak, A. Sidhu, A.; *J. Med. Chem.* 1, **2006,** 256.

27. M. Coghlan, A. Culbert, D. Cross, S. Corcoran, J. Yates, N. Pearce, O. Rausch, G. Murphy, P. Carter, L. Cox, D. Mills, M. Brown, D. Haigh, R. Ward, D. Smith, K. Murphy, A. Reith, J. Holder, *Chemistry and Biology,* 7, **2000,** 793.

28. D. L. Pavia, G. M. Lampman and G. S. Kriz, *Introduction to Spectroscopy,* 3rd Ed. ISBN 0030319617.

29. L. D. S. Yadav "*Organic Spectroscopy*" Anamaya Publisher, ISBN 1081-88342-41-6, New Delhi, **2005.**

30. A. Alparone; V. Librando, Spectrochimica Acta Part A: *Mole and Bimole. Spectro.* Vol. 89, xxx, **2012**.

31. J. Jayabharathi, V. Thanikachalam, M.V. Perumal Spectrochimica Acta Part A: *Molecular and Bimolecular Spectroscopy* Vol. 85, **2012**, 31-37.

32. L. Sinha, O. Prasad, V. Narayan, R. K. Srivastava, *Journal of Molecular Structure: Theochem* 23 July, **2010**.

33. V. Arjunan, S. Sakiladevi, T. Rani, C. V. Mythili, S. Mohan., *Spectrochimica Acta* Part A: *Molecular and Bimolecular Spectroscopy* Vol. 88, **2012**, 220-231.

34. R. A. Friesner, B. D. Dunietz, *Acc Chem Res* Vol. 34 (5), **2001**, 351-358.

35. W. Thiel, *Semi Empirical quantum chemical methods in Computational Chemistry, Elsevier* Chapter 21, **2005**, 559-580.

36. R. B. Murphy, D. M. Philipp, R. A. Friesner, *Journal of Computational Chemistry,* Vol. 21(16), **2000**, 1442-1457.

37. R. A. Friesner, V. Guallar *Annu Rev Phys Chem.* Vol. 56, **2005**, 389-427.

38. Exploring chemistry with Electronic Structure Methods, Second Ed^{n}. James B. Foresman and AEleen Frisch, *Gaussian, Inc.* Pittsburgh, PA.

39. A. Raymond, Serway and J. S. Faughn, College Physics" 2nd edition, University of James Madison. London.

40. M.A Rao, *Rheology of liquid foods, Journal Tenure studies,* **1997**, 135-168.

41. C. Klotutar, D. Rudan-Tasic and K. Romih, Densities and thermal expansion coefficients of some vegetable oils. *Acta chim. Slov.* **1997**, 45-55.

42. M. A. Eitman and J. W. Goodrum, *Rheology of the triglycerides tricaprion, tricaprylin, tricaprin and diesel fuel. Transactions ASAE,* **1993**, 503-507.

Chapter 2

Characterization of Synthesized Dihydropyrimidines and its Derivatives

2.1 Introduction

Compounds that contains nitrogen plays crucial and vital role in living things including plants, animals and human beings. These compounds throughout present in living environment of earth and plays an important metabolite present in all living organism. Today nearly, 70-80 per cent drugs are incorporated with amine functionality [1]. Biginelli compounds *i.e.* 3,4 dihydropyrimidin-2(1H)-ones (DHPM's) possess biological applications [2-14]. In 1930, 4-chlorophenyl-2-thio- dihydropyrimidin-2(1H)-ones was registered for patent due to presence of protective property for hairs of sheep (wool) against insects and moths (Ertan *et al.*, 1933). Since from this time, attention has been adapted on antiviral activity of Biginelli compounds which resulted to discovery of outstanding antiviral agent nitractin (Ann and Hurst, 1962).

Some of the ancient believed "no reaction is possible without the use of a solvent" now is invalid. It is more prominent that many reactions are carried out on solid phase with no use of solvent. Infact many reactions are more efficient and selective than reactions done by using solvents. This types of reactions are simple, easy to handle, reduces environmental pollution, cheap to be utilized and has significance value in industry. It is also believed that solvent free organic synthesis industrially very useful and has green approach. But few of them requires polar acids, Lewis acids, elevated temperature and long lasting reaction time.

These conditions are not adopted in green chemistry. Ultimately to satisfy such need, such reactions are modified to improve the resultant yields. By overlooking through different literature surveys, it is very necessary perform such reactions which has outcomes of high yields, less chemicals utilization, short reaction time and mild reaction conditions. The replacement of environmentally and hazardous free solvents *i.e.* solvent free condition and easily and cheaply available catalyst are the innovative trends.

Pyrimidines are unsaturated cyclic amides which is an important building block in organic synthesis.

Pyrimidines have many advantages in pharmaceutical and medicinal chemistry. It also shows potential for the Cox-2 enzyme inhibition [14]. This can create a new horizon for treatment of anti-inflammatory activity.

Dihydropyrimidine carbonitrile showed the defensive antibacterial and antifungal activities [15-16], analgesics [17], and antihypertensive [18], anti-inflammatory [19], antitumor [20-22], calcium channel blocker [23], HIV reverse transcriptase [24-25], Pyrimidine carbonitriles are also effective against carcinogenesis on Raji cell [26].

Pyrimidines play a crucial role in many biochemical processes hence are considered to be important moiety in synthesis of drugs. Different effords were made in developing high efficient asymmetric Biginelli reaction due to wide range of biological activities and remarkable pharmacological properties. Most of them possess calcium channel blockers, antihypertensive agent, α-1a antagonists [27], HIV-gp-120-CD-4 inhibitor [28], anticancer [29], sedatives, hypnotics [30], anticonvulsant [31], CNS stimulant [32], antibacterial [33-35] and fungicides [36].

Biginelli (1983), from University of Florence firstly reported acid catalysed heating of ternary cyclocondensation of aromatic aldehydes, ethyl acetoacetate and urea in ethanol at reflux temperature. Product isolated on cooling was recognised to be 3, 4-dihydropyrimidin-2(1H)-ones. Biginelli reported this method to be superficial method of preparation of dihydropyrimidiones. Typical reaction involving use of aldehyde, β-keto esters and urea under strong acidic conditions and protic solvents to give dihydropyrimidones as shown below.

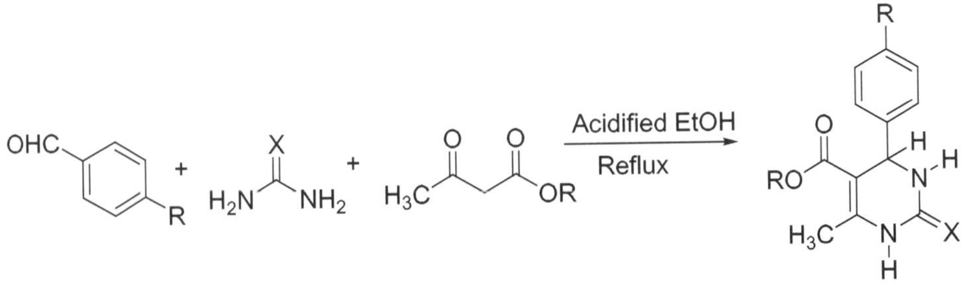

Biginelli reaction is considered as useful reaction which gives simple and effective path to achieve 3, 4-dihydropyrimidin-2(1H)-ones with different functional groups. The idea of generation of carbon-carbon double bond in single pot coupling reaction by using different components encourage to maintain high level of synthetically current method for molecular designing. By using different modes such as microwave irradiation, sonochemical and photochemical allows the reaction to be occur in fast rate in comparison to other traditional methods such as heating gives an increase in the yield of product. Furthermost, reactions which are temperature sensitive *i.e.* having thermodynamic or kinetic pathways can be adjusted and extracted selectively. As multi component reactions generally create complete molecular product in one step, it is more easy and accurate to use this procedure.

There are reactive sites on the pyrimidine ring which exerts an influence on reactivity of other causing pyrimidine to be similar to typical cyclic olefin. Hence different methods have been adopted for the preparation of biological active pyrimidines and its deivatives.

2.2 Literature Methods for the Synthesis of Substituted Dihydropyrimidine Carbonitriles, Dihydropyrimidine-2(1H)-ones and Dihydropyrimidine Carboxylates

Synthesis of pyrimidines derivatives are usually carried out by one step condensation procedure. According to us, the following are some of the procedures from literatures which are useful in preparation of dihydropyrimidines and its derivatives. The synthesis of dihydropyrimidine carbonitriles is started by condensation of aromatic aldehyde and ethyl cyanoacetate by nucleophilic attack of carbanion generated by weak base to produce α,β unsaturated ester. Further on reaction with urea/thiourea, elimination of alcohol and hydrogen molecule takes place to give product.

The work of synthesis of pyrimidine was started in 1979 by S. Kambe, K. Saito and H. Kishi [37] by ternary condensation of aromatic aldehyde, thiourea and ethylcyanoacetate in ethanol using potassium carbonate as a base for 8-10 hours.

A. Patel and co-workers [38] reported the synthesis by ternary condensation of aromatic aldehyde, thiourea and ethylcyanoacetate using potassium hydroxide in presence of dry methanol at 60-70°C which requires 12 hrs to complete the reaction.

Y. L. N Murthy and co-worker [39] reported the synthesis of pyrimidine carbonitrile by using magnesium methoxide in ethanol.

A. F Eveas and co-workers [40] synthesized pyrimidine carbonitriles by condensation of aromatic aldehyde, thiourea and ethylcyanoacetate using sodium ethoxide.

T. S. Chitre *et al.* [41] reported synthesis of pyrimidine carbonitriles using microwave and ultrasound radiation using commercially available potassium carbonate in presence of ethanol for 10-20 min.

M. S. Muftah [42] synthesized pyrimidine carbonirtiles by using piperidine and refluxing for 12 hrs.

The synthesis of pyrimidine carbonitrile in relative high yield was reported by S. A. Desai *et al.* [43] by using microwave and ultrasonic radiation for 7-8 min.

Kappe and Stadler [44] reported synthesis of dihydropryimidines by using microwave radiation and $Yb(OTf)_3$ as a catalyst.

Debache *et al.* [45] demonstrated synthesis of dihydropyrimidones using phenyl bororonic acid [$phB(OH)_2$] as a catalyst for synthesis of three component condensation of Biginelli reaction. The yield of the product was considerably high with diversed aromatic aldehydes.

P. V. Anbhule and co-workers [46] simply reported one step synthesis of 6-amino-5-cyano-4-phenyl-2-mercaptopyrimidine and its hydroxyl analogues.

J. C. Bussolari and P. A. McDonnell [47] reported the synthesis of dihydropyrimidones by direct preparation from β-keto carboxylic acids in absolute alcohol and refluxing for 12 hrs to give excellent yield of 75-90 per cent.

S. Patil and co-workers [48] used pineapple juice as a natural catalyst for Biginelli reaction for preparation of 3,4 dihydropyrimidine-2(1H)-ones by stirring at room temperature. The product was obtained with high yield.

B. Gangadasu and co-workers [49] used calcium chloride for synthesis of 3,4 dihydropyrimidine-2(1H)-ones in ethanol. The reaction completed in 2-6 hrs and gives product with excellent yield.

Y. Zhao *et al.* [50] used poyphosphoric acid for one pot synthesis of 3,4 dihydropyrimidine-2(1H)-ones by grinding under solvent free condition and room temperature to afford product with yield of 80-96 per cent.

N. D. Moirangthem and W. S. Laitonjam [51] used chromium (III) chloride for synthesis of pyrimidine-2-thiones in good yield.

M. B. Deshmukh and co-workers [52] used easily available phosphorus pentaoxide as efficient catalyst for synthesis of dihydropyrimidine-2(1H)-ones. The reaction mixture was reflux for one hour in water bath to afford product with good yield.

M. A. Pasha *et al.* [53] explored zinc chloride as a catalyst for Biginelli reaction using microwave radiation under solvent free condition with reaction time of 20-35 sec only.

G.Aridoss and Y. T. Jeong [54] used $Y(OAc)_3$ in acetic acid as a simple convenient reaction condition for synthesis of 3,4 dihydropyrimidine-2(1H)-ones.

M. R. Mohammadizadeh and N. Firoozi [55] used trifluoroacetic acid as an effective catalyst for one pot three component Biginelli reaction which yields 3, 4 dihydropyrimidine-2(1H)-ones and thiones in desirable amount.

T. Boumoud and co-workers [56] used magnesium nitrate as cheap and novel catalyst for one pot synthesis of substituted 3, 4 dihydropyrimidine-2(1H)-ones under solvent free condition.

A. Shaabani *et al.* [57] exploited ammonium chloride as a catalyst for the synthesis of 3,4 dihydropyrimidine-2(1H)-ones under solvent free condition by refluxing at 100°C for 2-3 hours.

P. Salehi and co-workers [58] used silica sulphuric acis as an effective and reusable calatyst for the one pot synthesis of 3,4 dihydropyrimidine-2(1H)-ones.

Garima *et al.* [59] started Biginelli reaction directly from aromatic alcohols by using Bronsted acidic ionic liquid [Hmim]HSO_4-NO_3 for 2-4 hrs at 80°C.

K. K. Pasunooti and co-workers [60] reported copper catalysed microwave assisted three component synthesis of dihydropyrimidones under mild condition.

S. V. Rami and co-workers [61] explored use of trifluoro ethanol as metal free, homogeneous and recyclable medium for efficient one pot synthesis of dihydropyrimidines.

Z. L. Shen and co-workers [62] used potassium ter.butoxide (tbuOK) as bronsted base catalysed one pot three component Biginelli type reaction.

Hence numerous methods are available in literature for the synthesis of dihydroprimidines and itd analogues [37-62] which contains single step condensation of aromatic aldehydes with urea or thiourea, 2, 4 dicarbonyl compounds to give corresponding dihydropyrimidine derivatives [50, 51, 54] as shown in Table 2.1

Table 2.1: Preparation of Dihydropyrimidine Carbonitrile: One Step Procedure

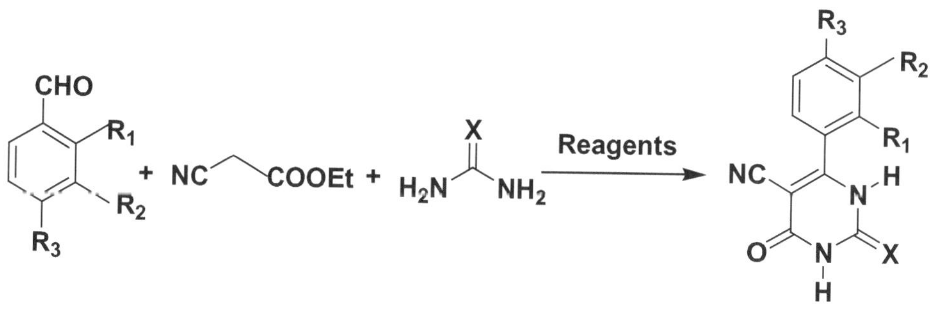

1 2 3 4

Sl.No.	Reagents for One Step Procedure	Time	Percent Yield	References
1	Dry Methanol, KOH, 60-70°C.	12h	70-75	38
2	$Mg(OMe)_2$, Ethanol, reflux	5-8 h	75-85	39
3	C_2H_5ONa, stir	48h	90-96	40
4	K_2CO_3, Ethanol, Microwave	10-20 min	80-87	41
5	Piperidine, reflux	12 h	90-96	42
6	Ultrasound radiation	40-50 min	70-75	43

Table 2.2: Preparation of Dihydropyrimidine-2(1H)-ones/Thione

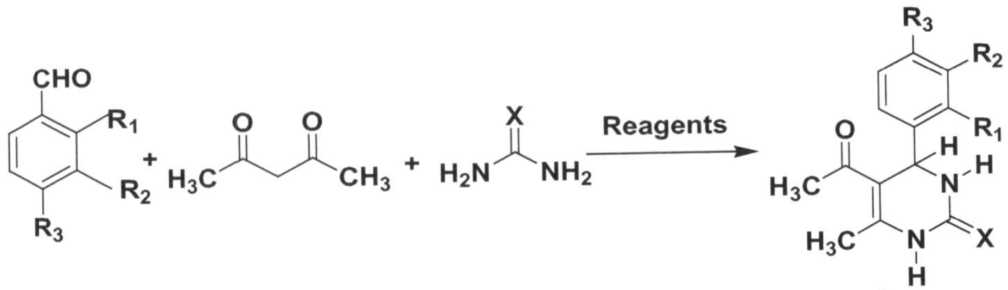

1 2 3 4

Sl.No.	*Reagents and Conditions*	*Time*	*Percent Yield*	*References*
1	Oxalactic acid,reflux	12 h	75.90	47
2	Pineapple juice, stir	2-4 h	70-90	48
3	Calcium Chloride, Ethanol, heat	2-6 h	80-98	49
4	Polyphosphoric acid, grinding, room temperature	25-30 min	85-93	50
5	Phosphorus pentaoxide, Ethanol, reflux	1-3.5 h	75-95	52
6	Zinc Chloride, microwave	20-35 sec	90-95	53
7	$Y(OAC)_3.xH_2O$, acetic acid, reflux	1-5 h	70-90	54

* Trifluoroacetic acid [55], $MgNO_3$ [56], Silica sulphuric acid[58], Copper [60]

Many of these methods involve long reaction time, anhydrous solvents, stoichiometric amount, hazardous radiations, use of costly apparatus and give unsatisfactory yields. Therefore in our present work we have used very less expensive and easily available catalyst as well as mild and neutral reaction conditions for the synthesis of 2,4 dihydro pyrimidine carbonitrile.

Hence we wish to report the results obtained from study of preparation of dihydropyrimidine carbonitrile and its derivatives with NH_4Cl as easily available catalyst under neutral and solvent free condition [63]. The procedure gives product in good yields and avoid the problems such as cost, handling and safety associated with the use of the solvent.

The method has decreased reaction time due to enhancing the reactivity of the reactant in solid state at 100°C reaction temperature.

2.3 Present Work

2.3.1 Synthesis of Dihydropyrimidine Carbonitrile (4a-4e)

Aromatic aldehyde (**1 a-e**, scheme-1), ethyl cyanoacetate, thiourea and NH_4Cl were heated under stirring at 100 °C for 3 hrs. After cooling, solid product obtained was filtered and washed with cold water. It was further recrystallized from ethanol or ethyl acetate: *n*-hexane (1:3) to have analytical sample which were characterized by IR and ^{1}HNMR spectroscopy.

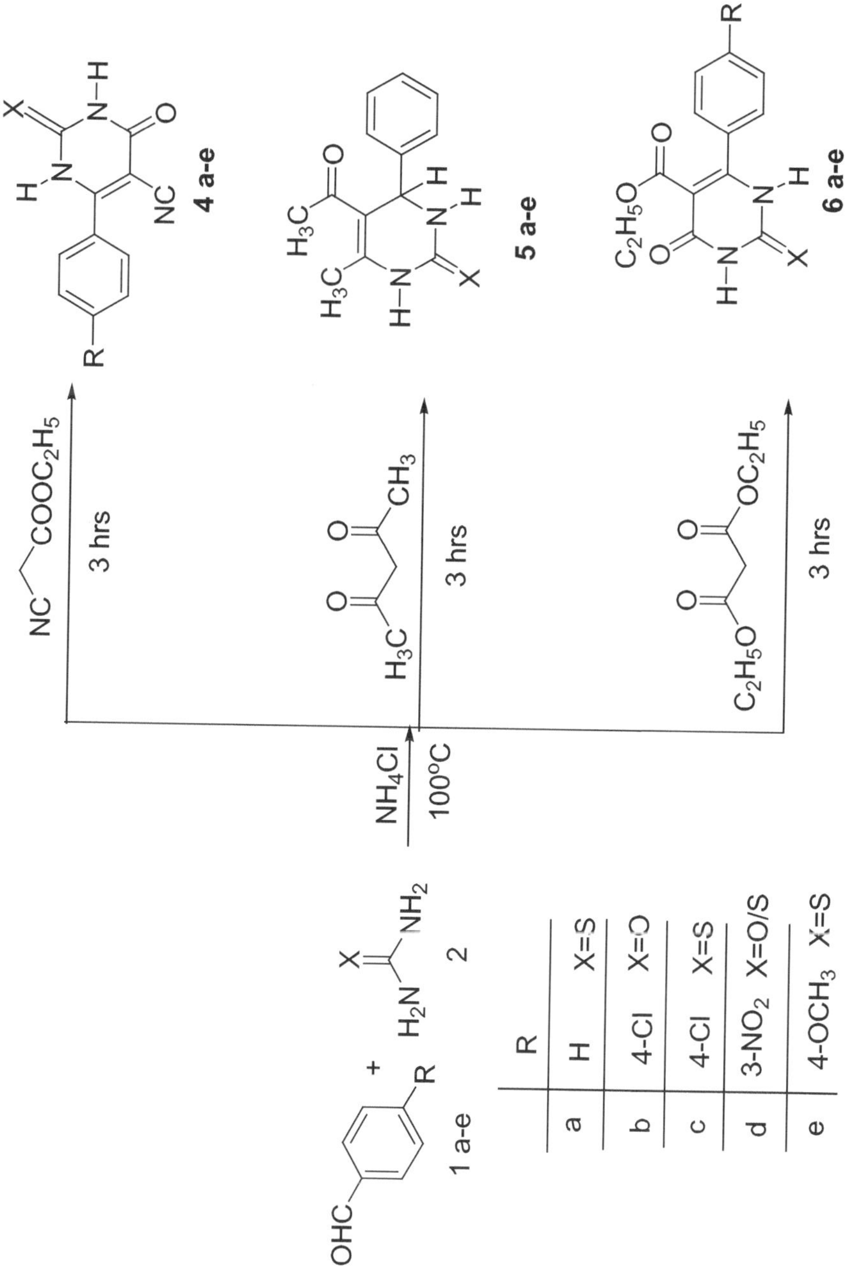
OHC
R
1 a-e
+
X
H2N NH2
2
NH4Cl
100°C
NC COOC2H5
3 hrs
H3C CH3
O O
3 hrs
C2H5O OC2H5
O O
3 hrs
4 a-e
5 a-e
6 a-e
R
a H X=S
b 4-Cl X=O
c 4-Cl X=S
d 3-NO2 X=O/S
e 4-OCH3 X=S

Scheme-1

IR Spectrum Data of Dihydropyrimidine Carbonitrile (4a)

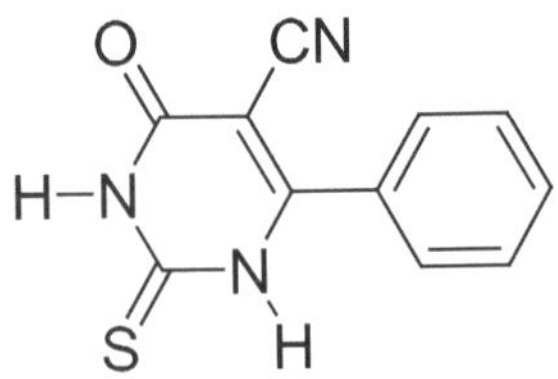

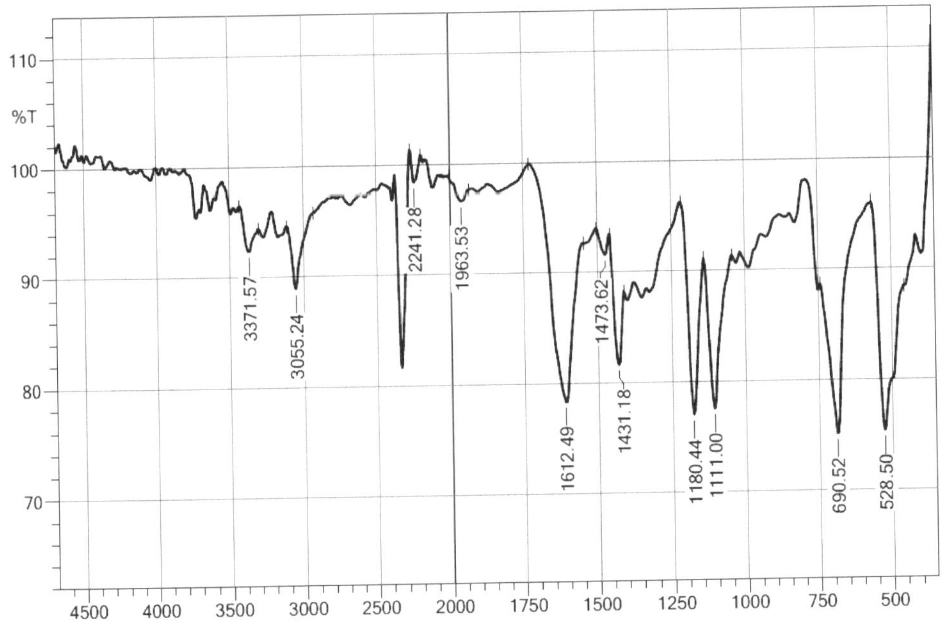

Figure 2.1: IR Spectrum of Dihydropyrimidine Carbonitrile (4a)

Infrared (IR) spectrum were recorded on Shimadzu FTIR spectrometer. In IR spectrum of 4a cyano frequency is seen at 2241 cm^{-1} and 1680 cm^{-1} confirms the -NH and –C=O from amide group. Absorption at 3371 cm^{-1} indicates -N-H group and 1180 cm^{-1}confirms presence of C=S group.

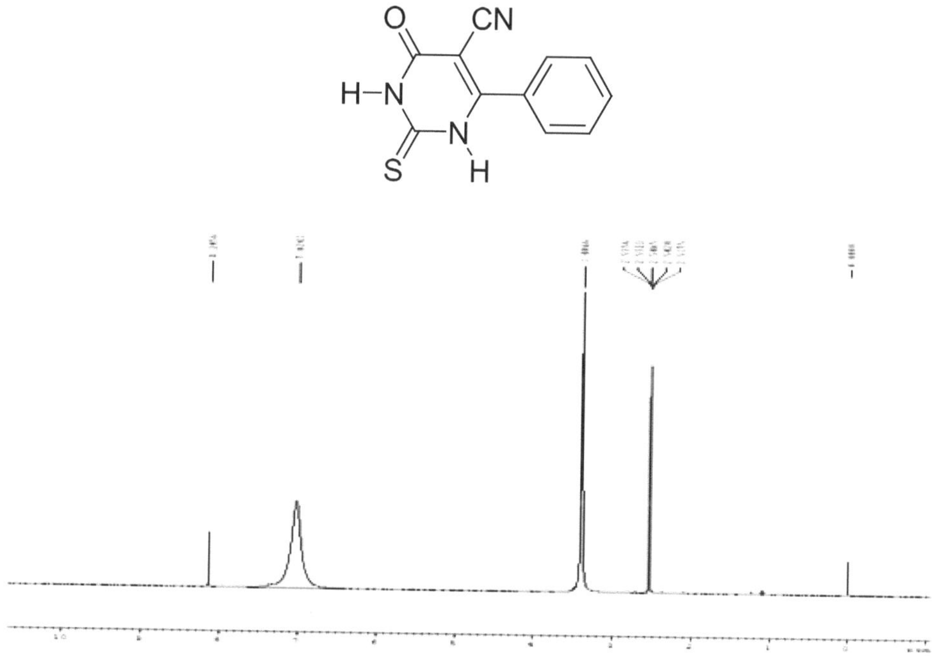

Figure 2.2: ¹H-NMR Spectrum of Dihydropyrimidine Carbonitrile (4a).

¹H-NMR spectrum: ¹HNMR Spectra were recorded on Bruker Advance II (400 MHz) using DMSO-d_6 solvent. Aromatic protons appeared in the range 7.6-8.0 δ while the NH proton gives singlet peak at 8.1 δ.

2.3.2 Synthesis of Dihyropyrimidin-2(1H)-one (5a-5e)

Aromatic aldehyde (**1 a-e**, scheme-1), acetyl acetone, thiourea and NH_4Cl was heated under stirring at 100 °C for 3 hrs. After cooling, solid product obtained was filtered and washed with cold water. It was further recrystallized from ethanol or ethyl acetate: *n*-hexane (1:3) to have analytical sample which were characterized by IR and ¹H-NMR spectroscopy.

IR Spectrum data of dihyropyrimidine-2(1H)-one (5a)

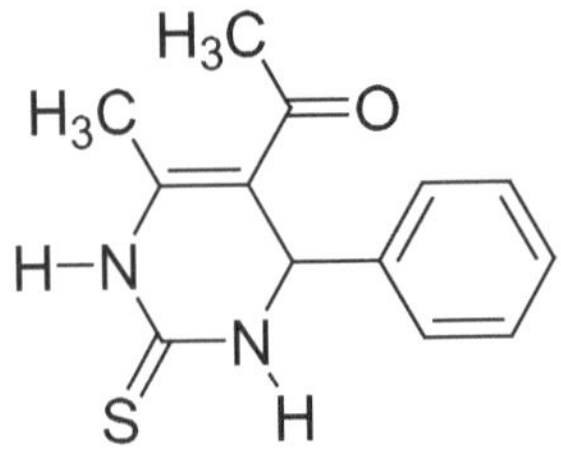

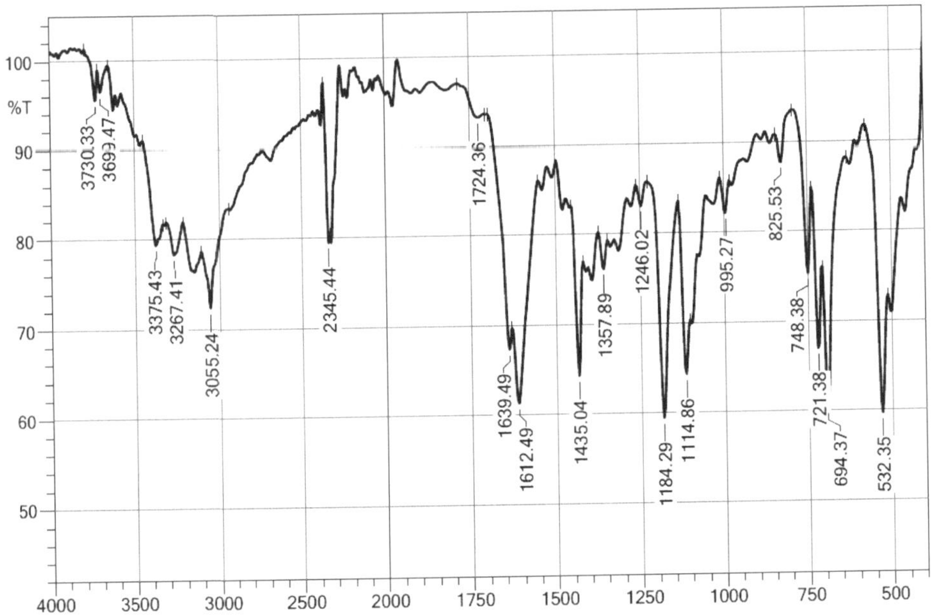

Figure 2.3: IR Spectrum of Dihyropyrimidin-2(1H)-one (5a).

IR spectrum (5a): In the IR spectrum of 5a both the carbonyl frequencies were observed at 1640 cm^{-1} and 1724 cm^{-1} corresponding to amide carbonyl and other carbonyl of ketonic group respectively. The olefinic C=C bond stretching appeared at 3267 cm^{-1}. While aromatic C-H stretching was seen at 3267 cm^{-1}. The N-H frequency is seen at 3730 cm^{-1}. The C-H stretching in methyl group is observed at 3055 cm^{-1} while absorption observed at 1114 cm^{-1} is due to presence of C=S stretching.

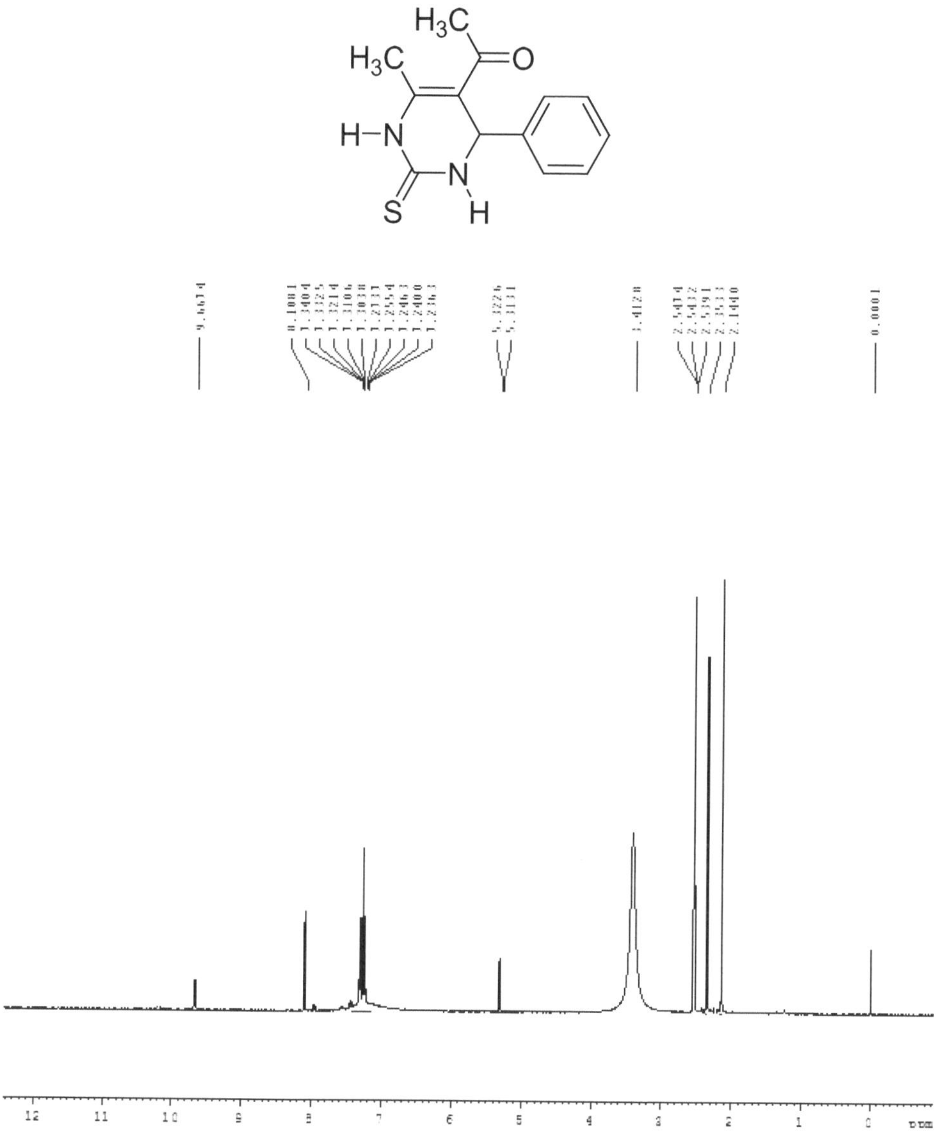

Figure 2.4: ^{1}H NMR Spectrum of Dihyropyrimidin-2(1H)-one (5a).

^{1}H-NMR spectrum (5a): Compound 5a was predicted from spectrum of ^{1}H-NMR spectroscopy in which hydrogen's from two methyl appears at 2.1 and 2.35 δ respectively. One protons of methine group appear at 5.32 δ. Two N-H protons appear at 8.1 and 9.66 δ as singlet. Aromatic protons give multiplet at 7.23 to 7.34 δ respectively.

2.3.3 Synthesis of Dihydropyrimidine Carboxylates (6a-6e)

Aromatic aldehyde (**1 a-e**, scheme-1), diethyl malonate, thiourea and NH_4Cl were heated under stirring at 100 °C for 3 hrs. After cooling, solid product obtained was filtered and washed with cold water. It was further recrystallized from ethanol or ethyl acetate: *n*-hexane (1:3) to have analytical sample which were characterized by IR and ^{1}HNMR spectroscopy.

Spectral Data of Dihydropyrimidine Carboxylate (6a)

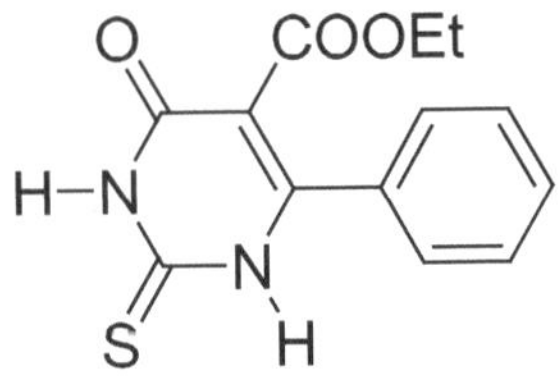

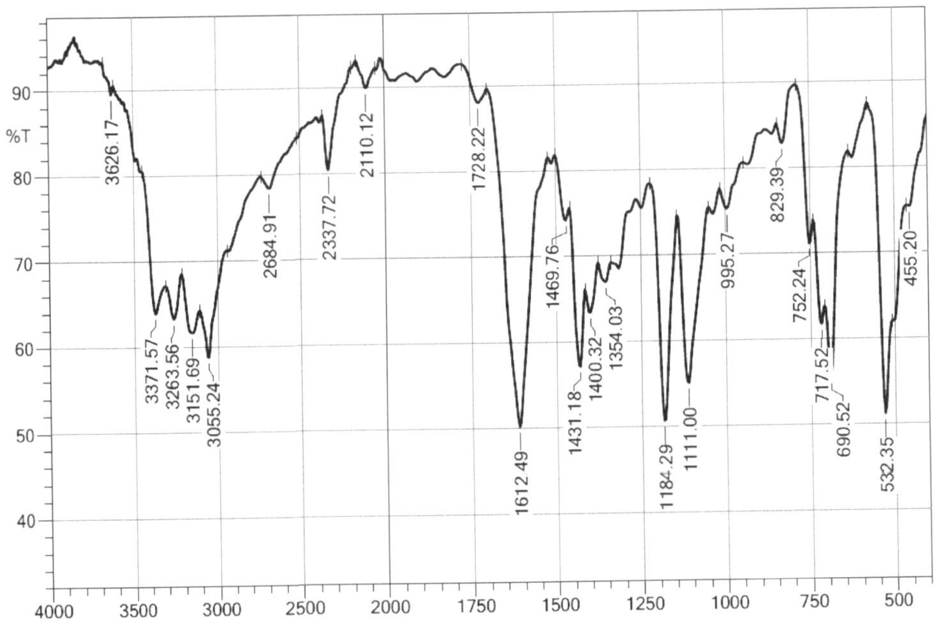

Figure 2.5: IR Spectrum of Dihydropyrimidine Carboxylates (6a).

IR spectrum (6a): In the IR spectrum of 6a carbonyl frequency was observed at 1728 cm^{-1} confirms the presence of carbonyl from carboxylic group. Aromatic C-H stretching was seen at 3263 cm^{-1}. Stretching frequency of amide –NH group appears at 3371 cm^{-1} while the stretching frequency of C-O in ester group was observed at 1111 cm^{-1}. Absorption at 1184 cm^{-1} corresponds to C=S stretching. Frequencies corresponding to 1612 cm^{-1} and 1728 cm^{-1} confirm presence of carbonyl group of amide and ester group respectively.

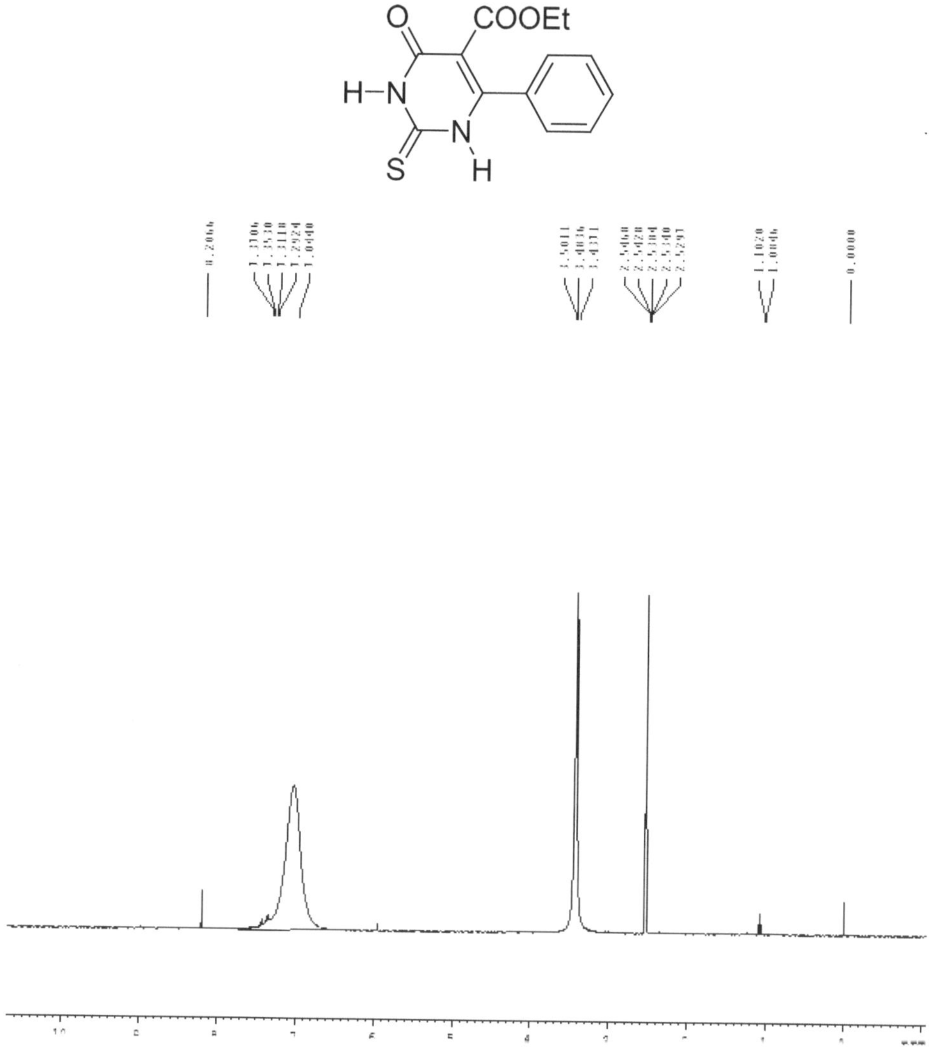

Figure 2.6: ^{1}H-NMR Spectrum of Dihydropyrimidine Carboxylates (6a).

^{1}H-NMR data (6a): The structure of the compound 6a was confirmed from 1H NMR spectrum. In which three protons from methyl group appeared at 1.10 δ. Two protons of O-CH_2 group appears at 3.50 δ. N-H protons appear at 8.26 δ. Aromatic protons gives multiplet at 7.0 to 7.35 δ respectively.

2.4 Experimental

2.4.1 General Procedure for Synthesis of Dihydropyrimidine Carbonitrile

A mixture of substituted aromatic aldehyde (0.30 g, 2 mmol), ethyl cyanoacetate (0.26 g, 2 mmole) and urea/thiourea were heated under stirring at 100 °C for 3 hrs. After cooling, solid product (**4a-e**) obtained was filtered and washed with cold water. It was further recrystallized from ethanol or ethyl acetate: *n*-hexane (1:3) to have analytical sample for spectral analysis.

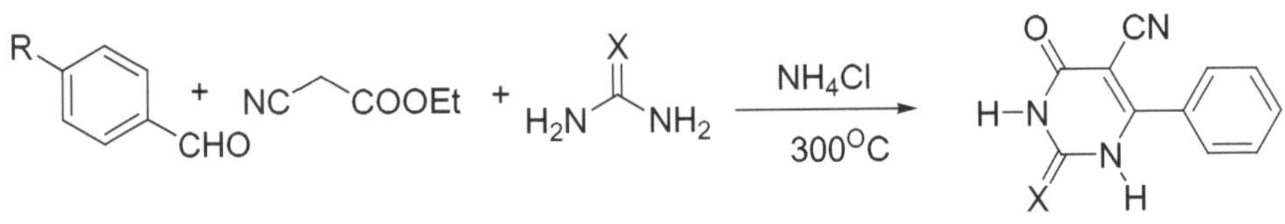

1 a-e 2 3 4 a-e

	a	b	c	d	e
R	H	4-Cl	4-Cl	$3\text{-}NO_2$	$4\text{-}OCH_3$

X= O/ S

Spectral Data

1,2,3,4-tetrahydro-4-oxo-6-phenyl-2-thioxopyrimidine-5-carbonitrile (4a) : Yield 75 per cent, m.p 296-298 °C, **IR (cm^{-1}):** 3372, 3055, 2241, 1964, 1613, 1474, 1431 cm^{-1}. **^{1}H NMR (DMSO-d_6 400 MHz):** 8.1 ∂ (s, -NH), 7.6-8.0 ∂ (m, 5 H, ArH). 6-(4-chlorophenyl)-1, 2, 3, 4-tetrahydro-2,4-dioxopyrimidine-5-carbonitrile (4b) : Yield 82 per cent, m.p 238-240 °C, **IR (cm^{-1}):** 3372, 3267, 3163, 2684, 2230, 2110, 1605, 1466, 1404, 1177, 1084, 725, 628, 498 cm^{-1}. **^{1}H NMR (DMSO-d_6 400 MHz):** 8.36 ∂ (s, -NH), 8.17 ∂ (s, -NH), 7.58 ∂ (d, 1 H, ArH), 7.60 ∂ (d, 1 H, ArH), 8.04 ∂ (d, 1 H, ArH) 8.05 ∂ (d, 1 H, ArH).

6-(4-chlorophenyl)-1, 2, 3, 4-tetrahydro—4-oxo-2-thioxopyrimidine-5-carbonitrile (4c): Yield 85 per cent, m.p 260-262 °C, **IR (cm^{-1}):** 3375, 3271, 3167, 2685, 2337, 2234, 1609, 1470, 1408, 1180, 1083, 725, 691, 628 cm^{-1}. **^{1}H NMR (DMSO-d_6 400 MHz):** 8.38 ∂ (s, -NH), 8.23 ∂ (s, -NH), 7.61 ∂ (d, 1 H, ArH), 7.63 ∂ (d, 1 H, ArH), 8.05 ∂ (d, 1 H, ArH), 8.07 ∂ (d, 1 H, ArH).

1, 2, 3, 4 tetrahydro-6-(3-nitrophenyl)-4-oxo-2-thioxopyrimidine-5-carbonitrile (4d) : Yield 72 per cent, m.p 275-278 °C, **IR (cm^{-1}):** 3375, 3271, 3167, 2684, 2338, 2114, 1609, 1470, 1408, 1246, 1180, 1114, 1088 cm^{-1}. **^{1}HNMR (DMSO-d_6 400 MHz):** 8.92 ∂ (s, -NH), 8.51 ∂ (s, -NH), 8.44 ∂ (d, 1 H, ArH), 8.42 ∂ (d, 1 H, ArH), 7.88 ∂ (d, 1 H, ArH), 7.84 ∂ (d, 1 H, ArH).

1,2,3,4 tetrahydro-6-(4-methoxyphenyl)-4-oxo-2-thioxopyrimidine-5-carbonitrile (4e): Yield 90 per cent, m.p 272-274 °C, **IR (cm^{-1}):** 3356, 3267,

3163, 2689, 2342, 2110, 1667, 1601, 1465, 1404, 1250, 1157, 1084 cm^{-1}. **^{1}HNMR (DMSO-d_6 400 MHz):** 8.27 ∂ (s, -NH), 8.23 ∂ (s, -NH), 7.23 ∂ (d, 1 H, ArH), 7.14 ∂ (d, 1 H, ArH), 7.11 ∂ (d, 1 H, ArH), 7.07 ∂ (d, 1 H, ArH), 3.35 ∂ (s, 3H, $-OCH_3$).

2.4.2 General Procedure for Synthesis of Dihyropyrimidin-2(1H)-one

A mixture of substituted aromatic aldehyde (0.30 g, 2 mmol), acetyl acetone (0.26 g, 2 mmole) and urea/thiourea was heated under stirring at 100 °C for 3 hrs. After cooling, solid product (**5a-e**) obtained was filtered and washed with cold water. It was further recrystallized from ethanol or ethyl acetate: *n*-hexane (1:3) to have analytical sample for spectral analysis.

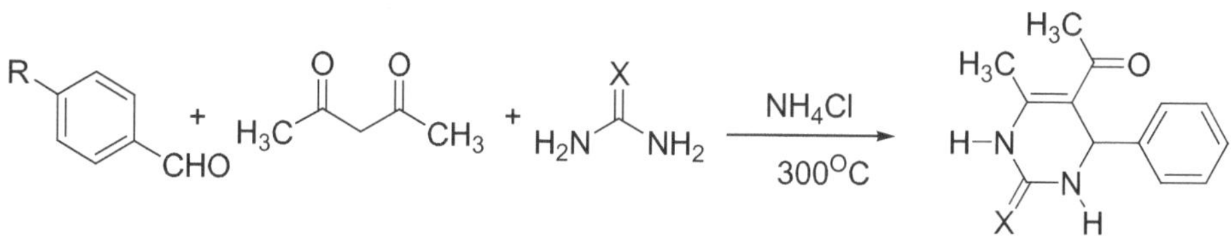

1 a-e 2 3 5 a-e

	a	b	c	d	e
R	H	4-Cl	4-Cl	3-NO_2	4-OCH_3

X= O/ S

Spectral Data

1-(1, 2, 3, 4-tetrahydro-6-methyl-4-phenyl-2-thioxopyrimidin-5-yl) ethanone (5a) : Yield 80 per cent, mp 220-222 °C, **IR (cm^{-1}):** 3730, 3699, 3375, 3267, 3055, 2345, 1724, 1639, 1612, 1435, 1358, 1246, 1184, 1115 cm^{-1}. **^{1} NMR (DMSO-d_6 400 MHz):** 9.66 ∂ (s, -NH), 8.11 ∂ (s, -NH), 7.23-7.34 ∂ (m, 5 H, ArH), 5.32 ∂ (s, 1 H, -CH), 2.35 ∂ (s, 3 H, $-COCH_3$), 1.10 ∂ (s, 3 H, $-CH_3$).

5-acetyl-4-(4-chlorophenyl)-3, 4-dihydro-6-methylpyrimidin-2(1H)-one (5b) : Yield 85 per cent, mp 208-210 °C, **IR (cm^{-1}):** 3371, 3271, 3167, 2684, 1609, 1469, 1411, 1180, 1084, 725, 629 cm^{-1}. **^{1}H NMR (DMSO-d_6 400 MHz) :** 9.23 ∂ (s, -NH), 8.15 ∂ (s, -NH), 7.48 ∂ (d, 1 H, ArH), 7.46 ∂ (d, 1 H, ArH), 7.44 ∂ (d, 1 H, ArH), 7.41 ∂ (d, 1 H, ArH), 5.67 ∂ (s, 1 H, -CH), 2.30 ∂ (s, 3 H, $-COCH_3$), 1.26 ∂ (s, 3 H, $-CH_3$).

1-[4-(4-chlorophenyl)-1, 2, 3, 4-tetrahydro-6-methyl-2-thioxopyrimidin-5-yl] ethanone (5c) : Yield 70 per cent, mp 220-222 °C, **IR (cm^{-1}):** 3371, 3271, 3167, 2689, 1608, 1470, 1411, 1180, 1084, 725, 502 cm^{-1}. **^{1}H NMR (DMSO-d_6 400 MHz) :** 9.70 ∂ (s, -NH), 8.14 ∂ (s, -NH), 7.39 ∂ (d, 1 H, ArH), 7.38 ∂ (d, 1 H, ArH), 7.35 ∂ (d, 1 H, ArH), 7.34 ∂ (d, 1 H, ArH), 5.33 ∂ (s, 1 H, -CH), 2.36 ∂ (s, 3 H, $-COCH_3$), 1.23 ∂ (s, 3H, $-CH_3$).

5-acetyl-3, 4-dihydro-6-methyl-4(3-nitrophenyl) pyrimidin-2(1H)-one (5d) : Yield 72 per cent, mp 230-232 °C, **IR (cm^{-1}):** 3375, 3271, 3167, 2685, 1728,

1608, 1469, 1411, 1246, 1180, 1118, 1088 cm^{-1}. **^{1}H NMR (DMSO-d_6 400 MHz) :** 9.32 ∂ (s, -NH), 8.28 ∂ (s, -NH), 7.69 ∂ (d, 1 H, ArH), 7.67 ∂ (d, 1 H, ArH), 7.64 ∂ (d, 1 H, ArH), 7.61 ∂ (d, 1 H, ArH), 5.42 ∂ (s, 1 H, -CH), 2.34 ∂ (s, 3 H, -$COCH_3$), 1.21 ∂ (s, 3 H, -CH_3).

1-(1,2,3,4-tetrahydro-4-(4-methoxyphenyl)-6-methyl-2-thioxopyrimidin-5-yl) ethanone (5e): Yield 80 per cent, mp 176-178 °C, **IR (cm^{-1}):** 3853, 3630, 3356, 3271, 3163, 2692, 1960, 1670, 1597, 1470, 1254, 1161, 1088 cm^{-1}. **^{1}H NMR (DMSO-d_6 400 MHz) :** 9.87 ∂ (s, -NH), 9.64 ∂ (s, -NH), 7.39 ∂ (d, 1 H, ArH), 7.26 ∂ (d, 1 H, ArH), 7.18 ∂ (d, 1 H, ArH), 7.16 ∂ (d, 1 H, ArH), 5.25 ∂ (s, 1 H, -CH), 3.46 ∂ (s, 3H, -OCH_3), 2.33 ∂ (s, 3 H, -$COCH_3$), 1.10 ∂ (s, 3 H, -CH_3).

2.4.3 General Procedure for Synthesis of Dihydropyrimidine Carboxylate

A mixture of substituted aromatic aldehyde (0.30 g, 2 mmol), diethyl malonate (0.26 g, 2 mmole) and urea/thiourea were heated under stirring at 100 °C for 3 hrs. After cooling, solid product (**6a-e**) obtained was filtered and washed with cold water. It was further recrystallized from ethanol or ethyl acetate: *n*-hexane (1:3) to have analytical sample for spectral analysis.

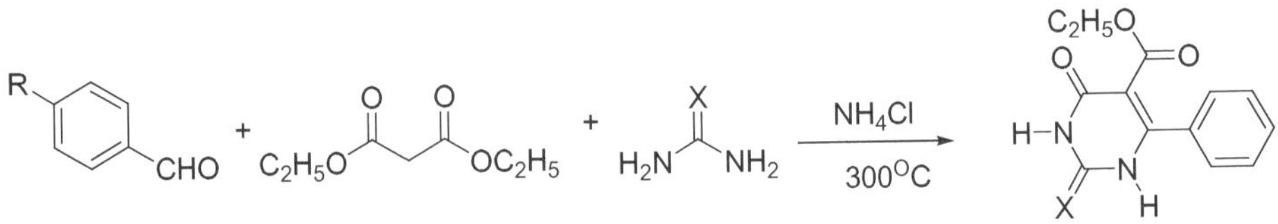

1 a-e 2 3 6 a-e

	a	b	c	d	e
R	H	4-Cl	4-Cl	3-NO_2	4-OCH_3

X= O/ S

Spectral Data

Ethyl 1, 2, 3, 4-tetrahydro-4-oxo-6-phenyl-2-thioxopyrimidine-5-carboxylate (6a) : Yield 70 per cent, mp 238-240 °C, **IR (cm^{-1}):** 3371, 3263, 3055, 2341, 1748, 1612, 1435, 1400, 1361, 1184, 1114 cm^{-1}. **^{1}H NMR (DMSO-d_6 400 MHz):** 8.2 ∂ (s, -NH), 7.04-7.37 ∂ (m, 5 H, ArH), 3.50 ∂ (q, OCH_2), 1.10 ∂ (t, -CH_3).

Ethyl 6-(4-chlorophenyl)-1, 2, 3, 4-tetrahydro-2,4-dioxopyrimidine-5-carboxylate (6b) : Yield 80 per cent, mp 240-242 °C, **IR (cm^{-1}):** 3367, 3163, 2337, 1760, 1605, 1465, 1408, 1183, 725, 494 cm^{-1}. **1H NMR (DMSO-d_6 400 MHz):** 8.15 ∂ (s, -NH), 7.42 ∂ (d, 1 H, ArH), 7.40 ∂ (d, 1 H, ArH), 7.38 ∂ (d, 1 H, ArH) 7.36 ∂ (d, 1 H, ArH), 3.61 ∂ (q, OCH_2), 1.16 ∂ (t, -CH_3).

2.6 IR and ^{1}H NMR Spectrum of Synthesized Deivatives

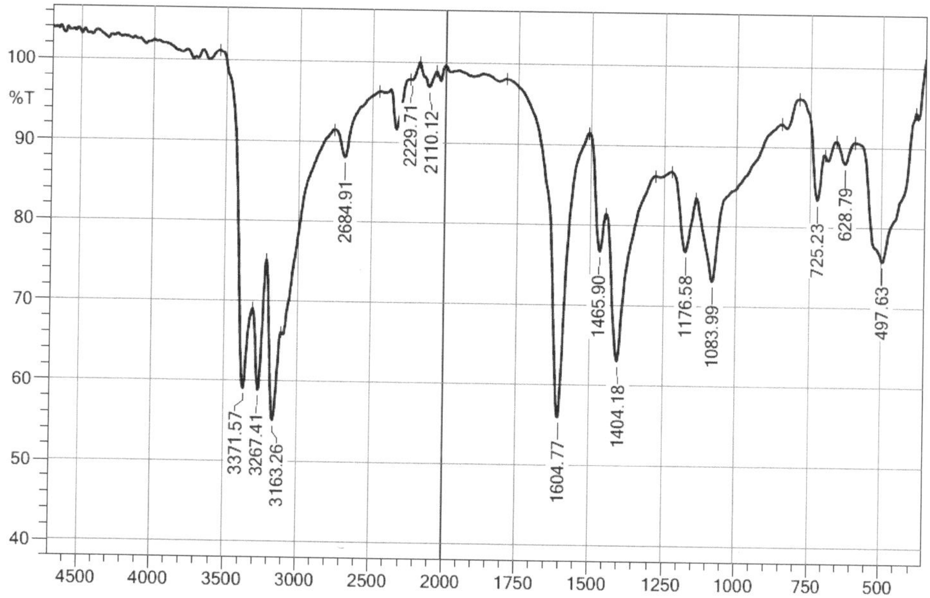

IR Spectrum (4b)

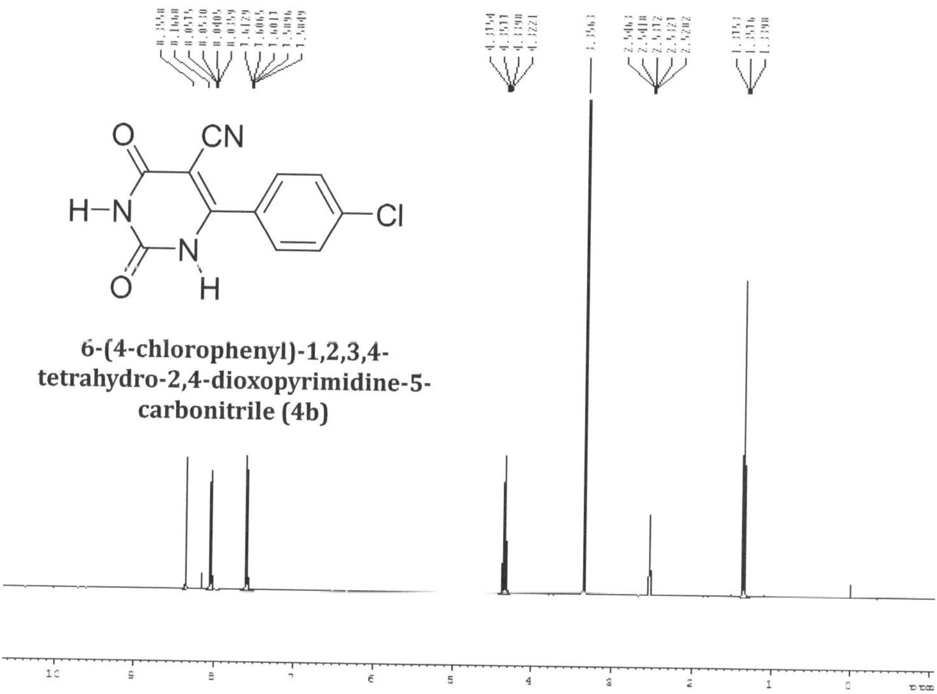

6-(4-chlorophenyl)-1,2,3,4-tetrahydro-2,4-dioxopyrimidine-5-carbonitrile (4b)

^{1}H-NMR Spectrum (4b)

Figure 2.8: Spectrum of Dihydropyrimidine Carbonitrile (4b).

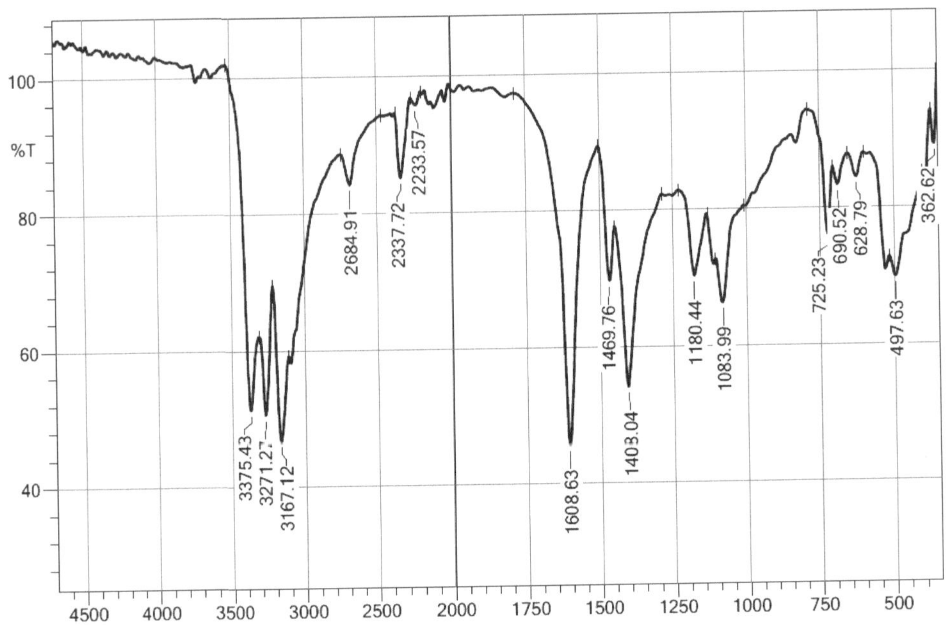

IR spectrum (4c)

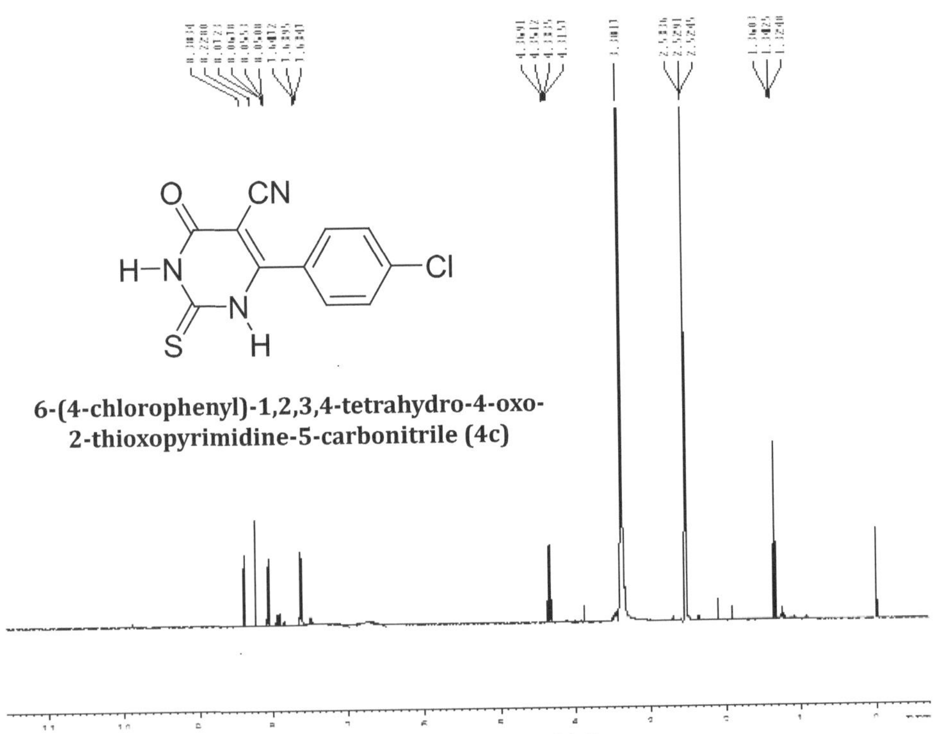

6-(4-chlorophenyl)-1,2,3,4-tetrahydro-4-oxo-2-thioxopyrimidine-5-carbonitrile (4c)

^{1}H-NMR spectrum (4c)

Figure 2.9: Spectrum of Dihydropyrimidine Carbonitrile (4c).

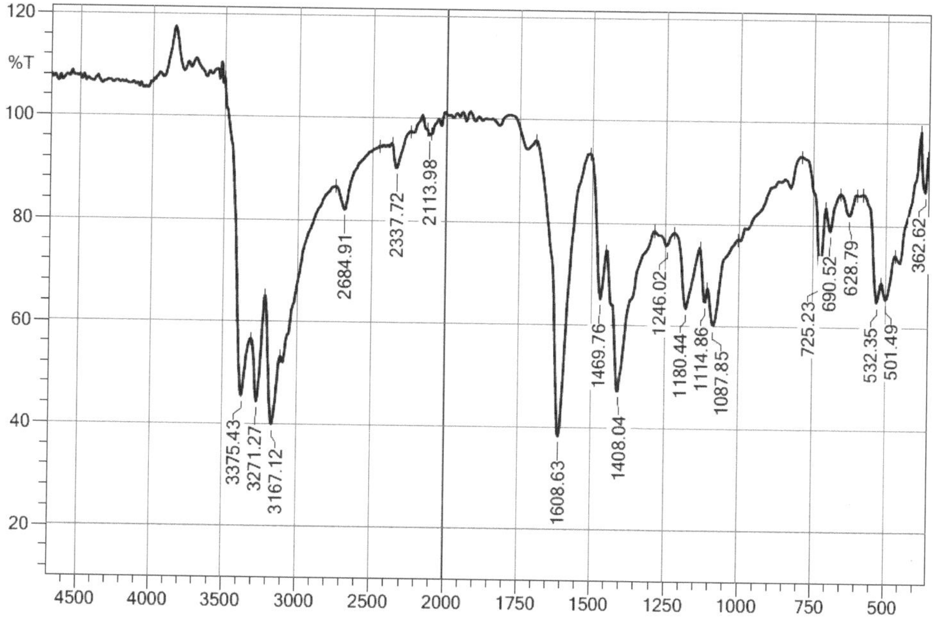

IR Spectrum (4d)

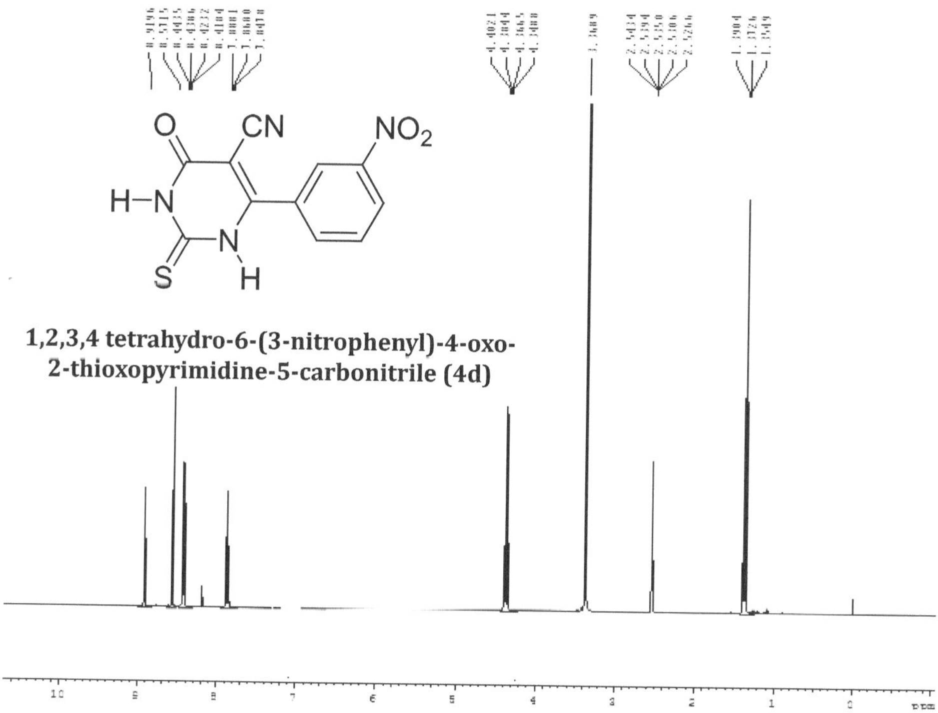

^{1}H-NMR spectrum (4d)

Figure 2.10: Spectrum of Dihydropyrimidine Carbonitrile (4d).

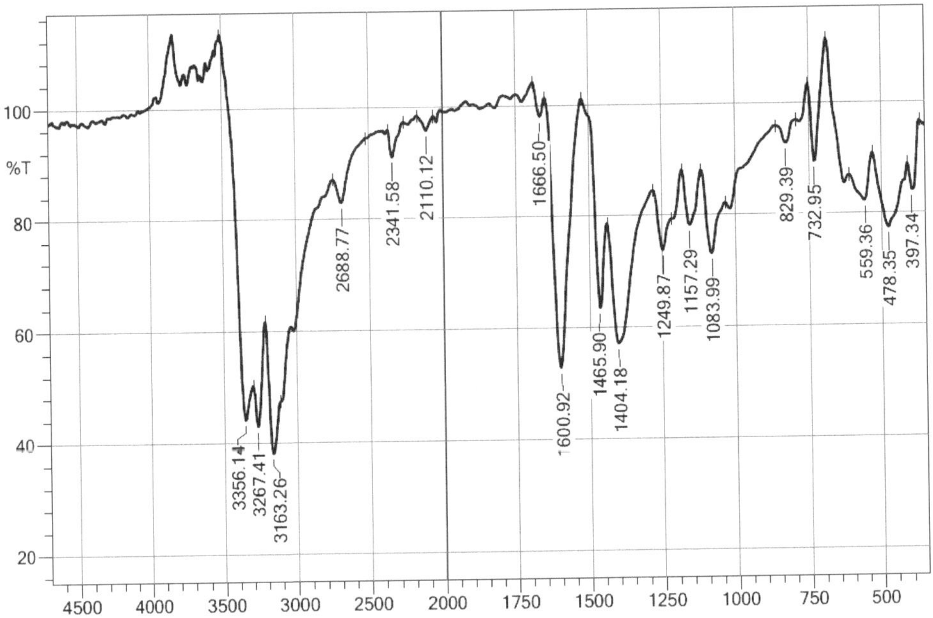

IR spectrum (4e)

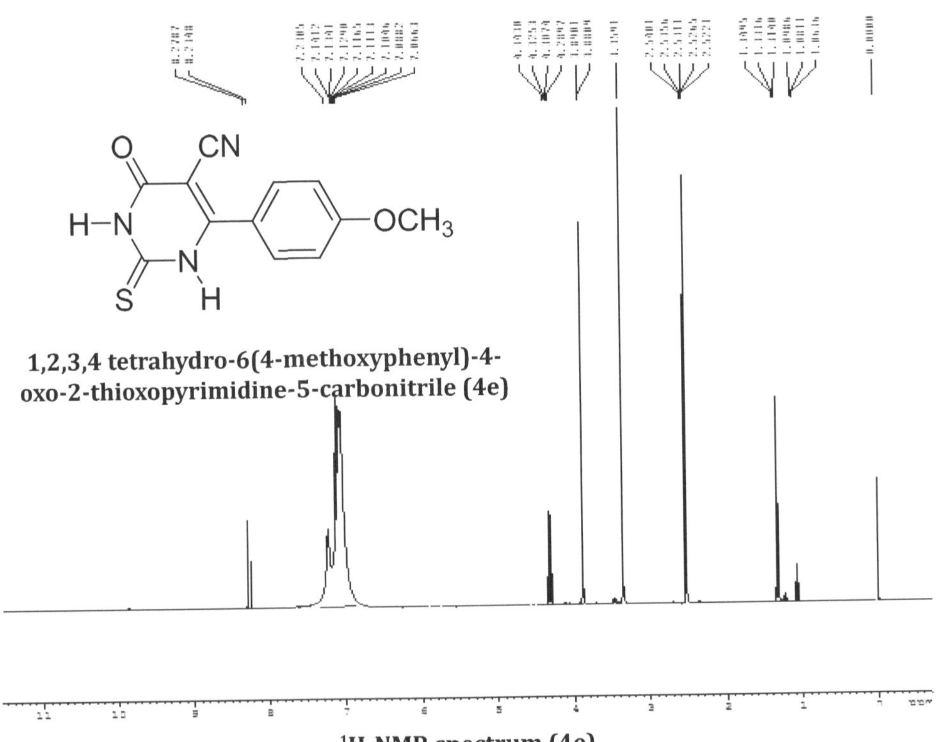

1,2,3,4 tetrahydro-6(4-methoxyphenyl)-4-oxo-2-thioxopyrimidine-5-carbonitrile (4e)

^{1}H-NMR spectrum (4e)

Figure 2.11: Spectrum of Dihydropyrimidine Carbonitrile (4e).

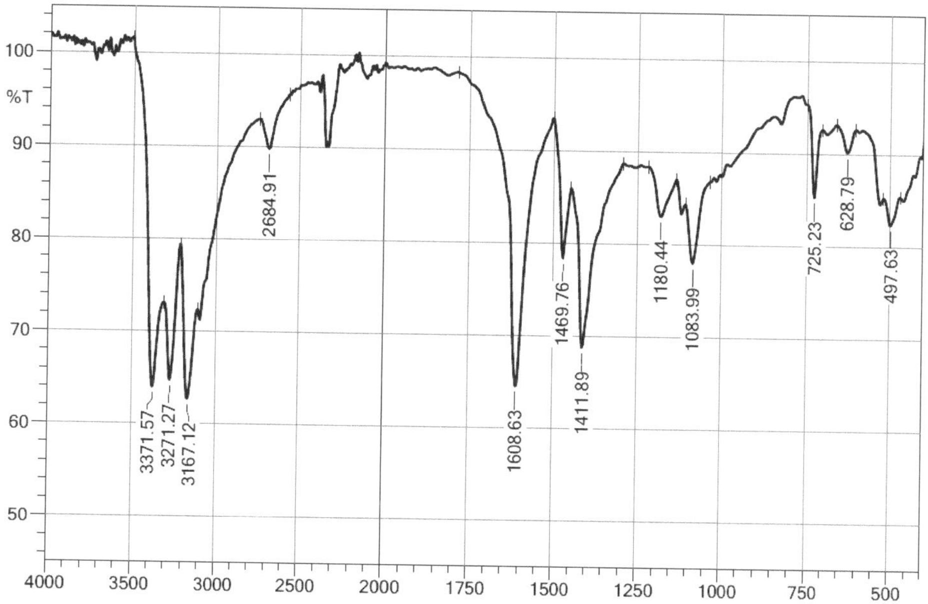

IR spectrum (5b)

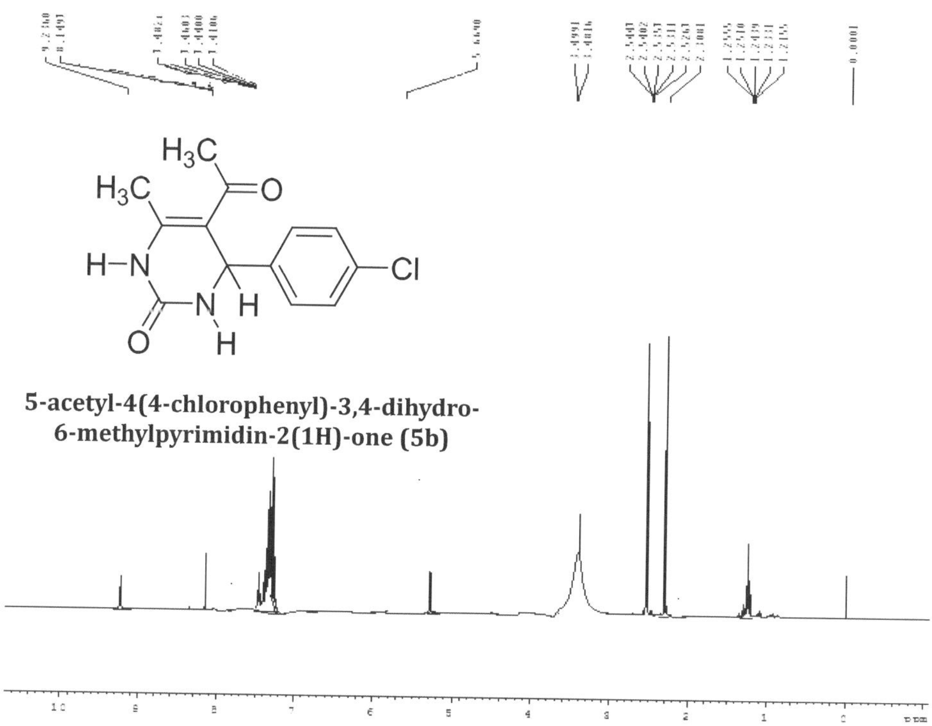

5-acetyl-4(4-chlorophenyl)-3,4-dihydro-6-methylpyrimidin-2(1H)-one (5b)

^{1}H-NMR spectrum (5b)

Figure 2.12: Spectrum of Dihyropyrimidin-2(1H)-one (5b).

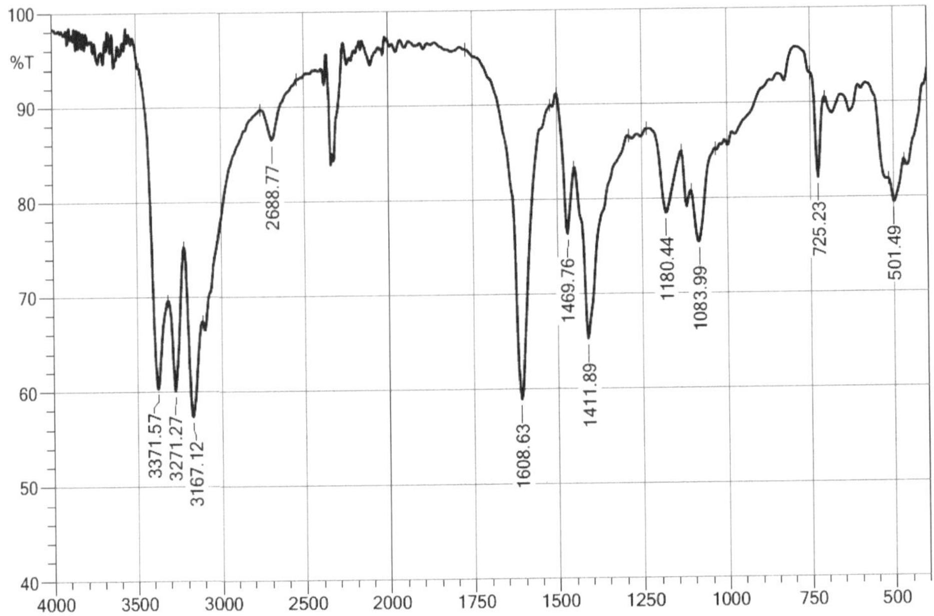

IR spectrum (5c)

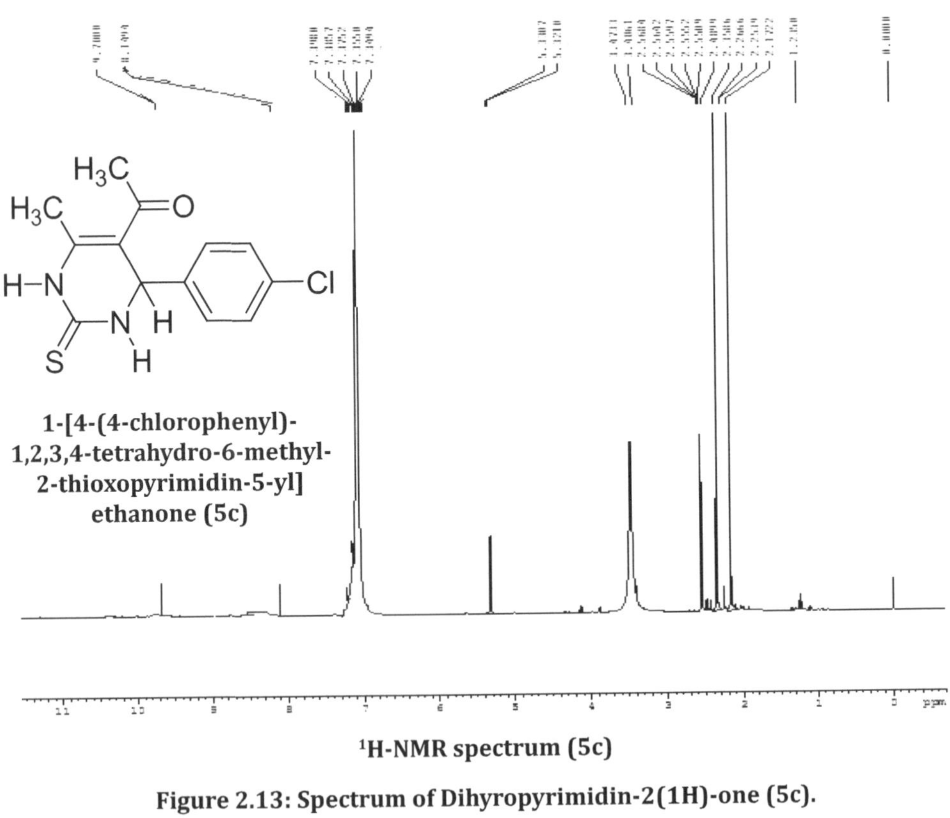

1-[4-(4-chlorophenyl)-1,2,3,4-tetrahydro-6-methyl-2-thioxopyrimidin-5-yl] ethanone (5c)

^{1}H-NMR spectrum (5c)

Figure 2.13: Spectrum of Dihyropyrimidin-2(1H)-one (5c).

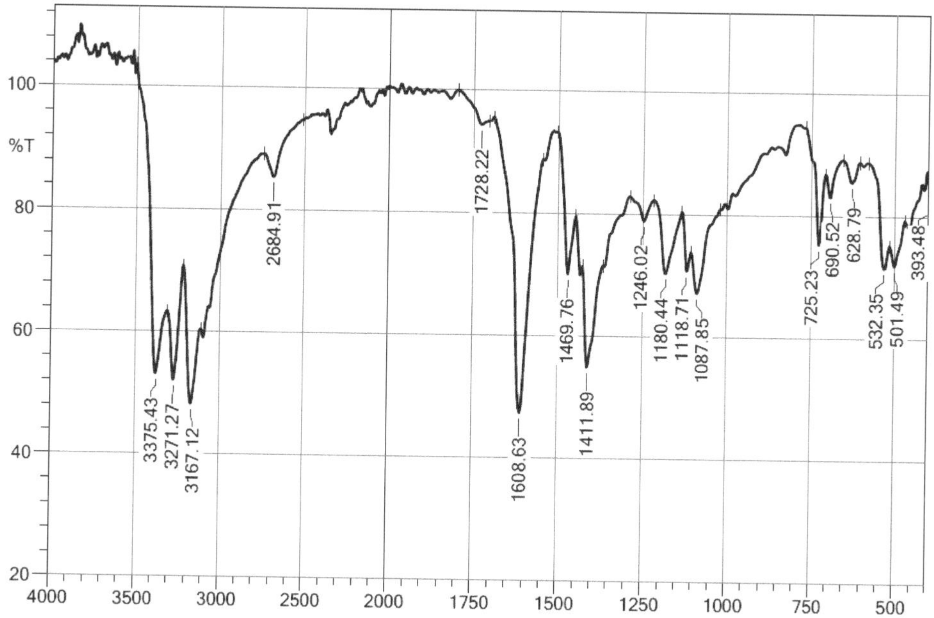

IR spectrum (4d)

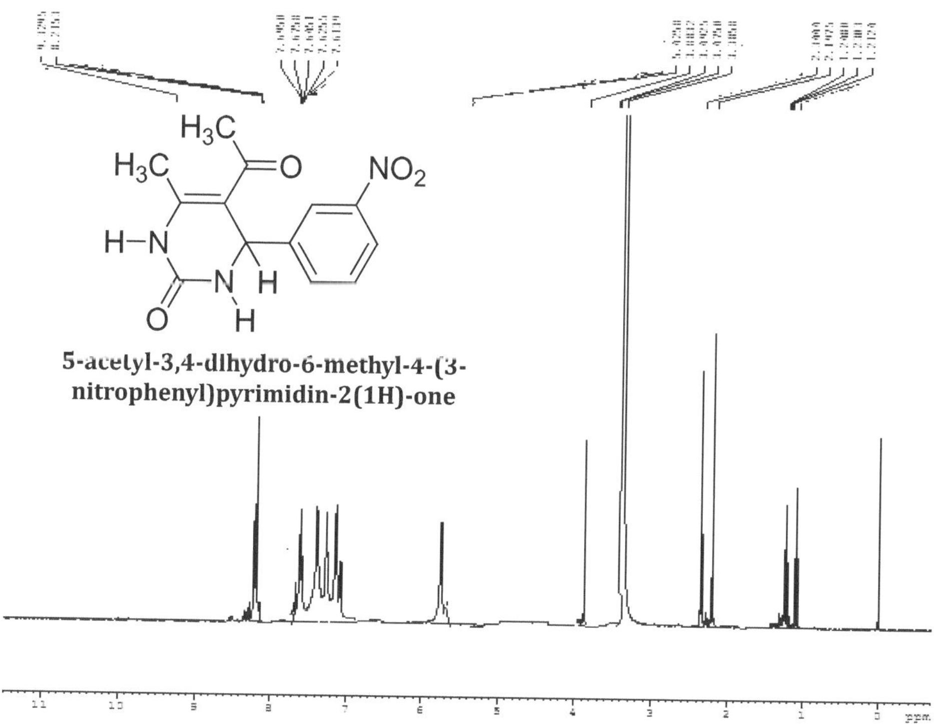

^{1}H-NMR spectrum (5d)

Figure 2.14: Spectrum of Dihyropyrimidine-2(1H)-one (5d).

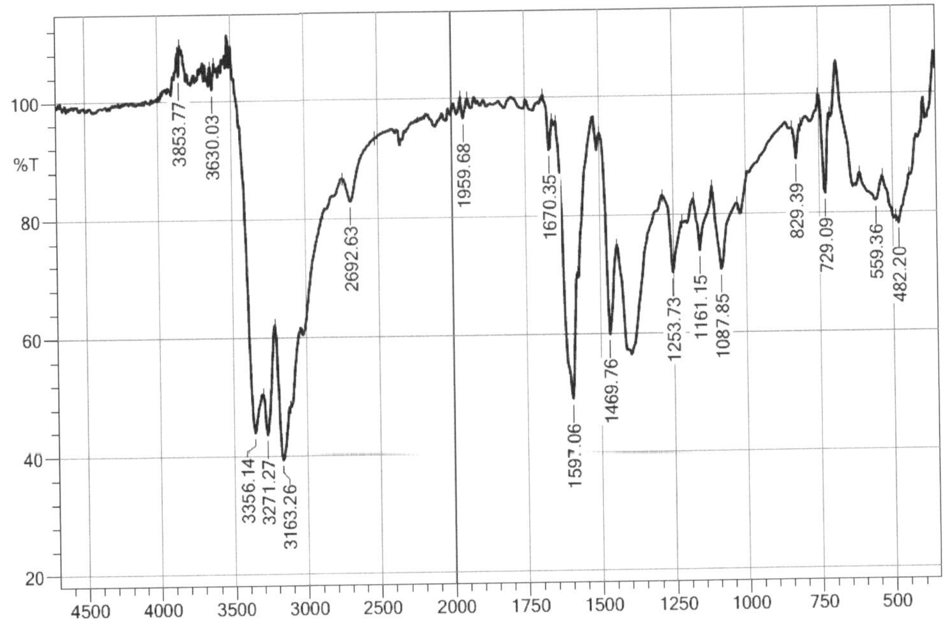

IR spectrum (5e)

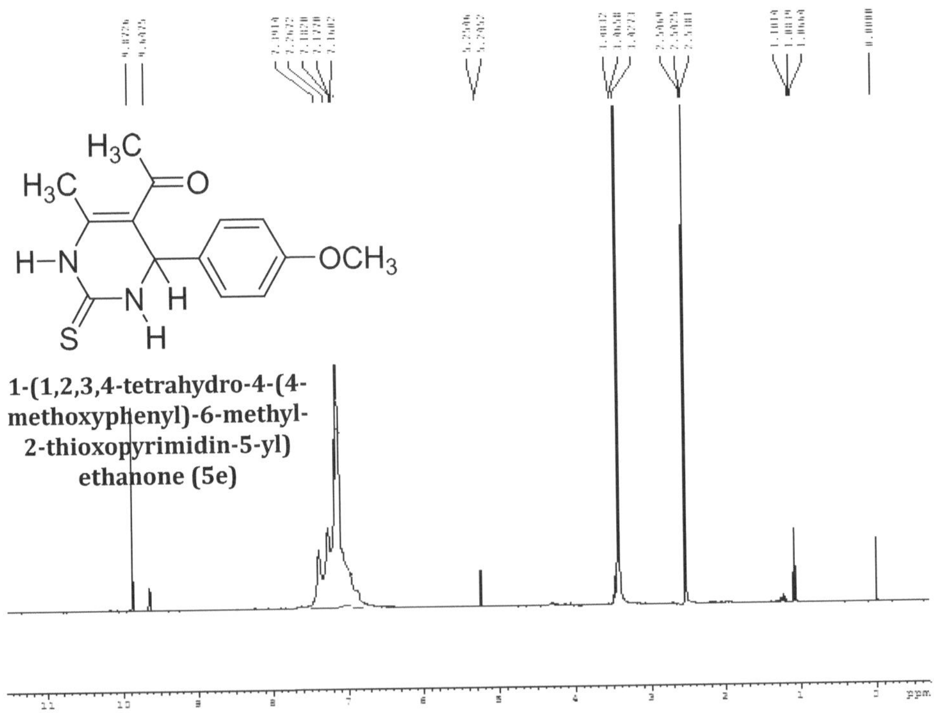

^{1}H-NMR spectrum (5e)

Figure 2.15: Spectrum of Dihyropyrimidine-2(1H)-one (5e).

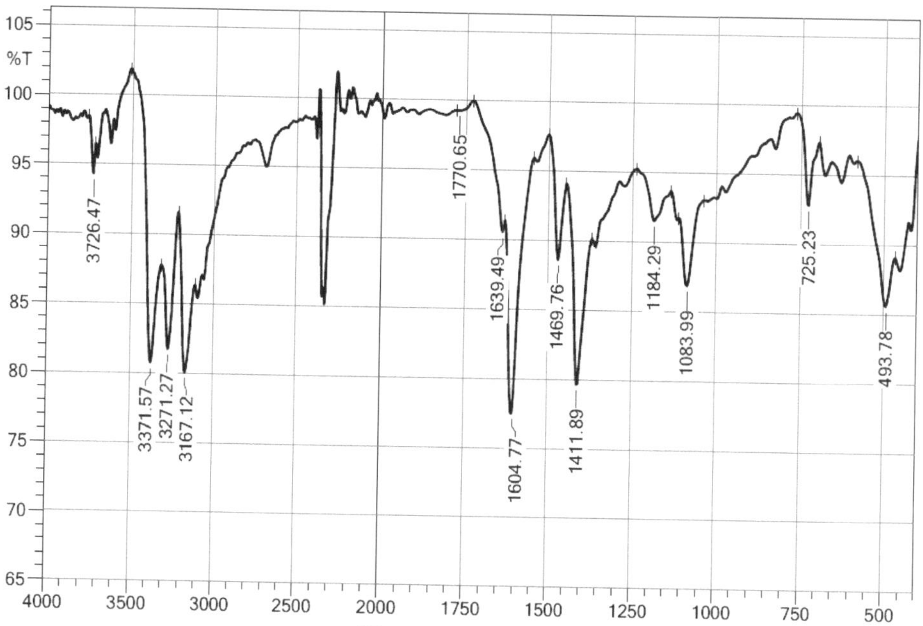

IR spectrum (6b)

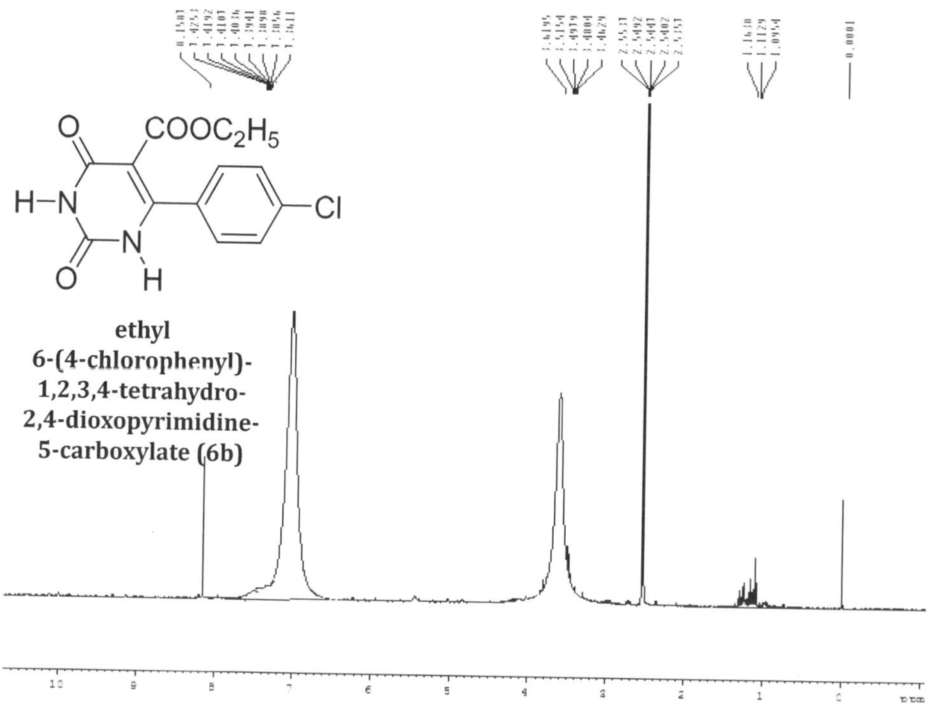

ethyl 6-(4-chlorophenyl)-1,2,3,4-tetrahydro-2,4-dioxopyrimidine-5-carboxylate (6b)

^{1}H-NMR spectrum (6b)

Figure 2.16: Spectrum of Dihydropyrimidine Carboxylate (6b).

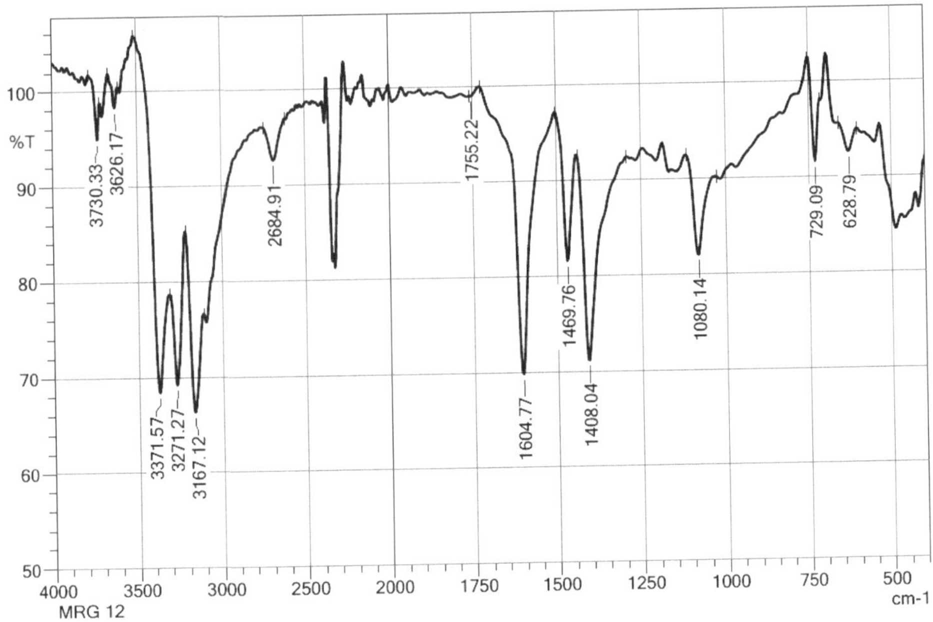

IR spectrum (6c)

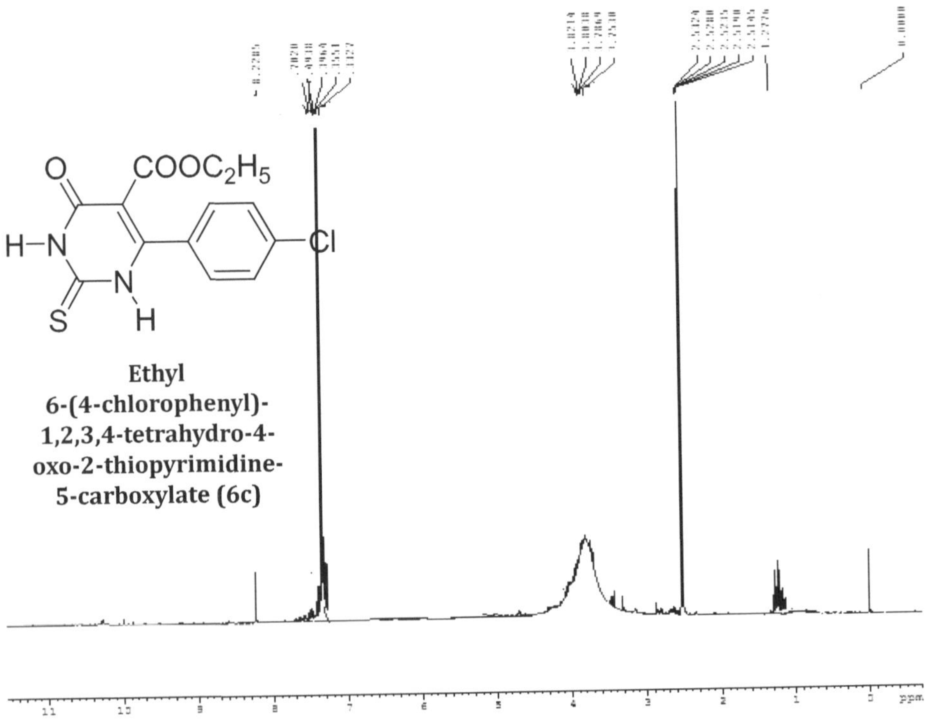

^{1}H-NMR spectrum (6c)

Figure 2.17: Spectrum of Dihydropyrimidine Carboxylate (6c).

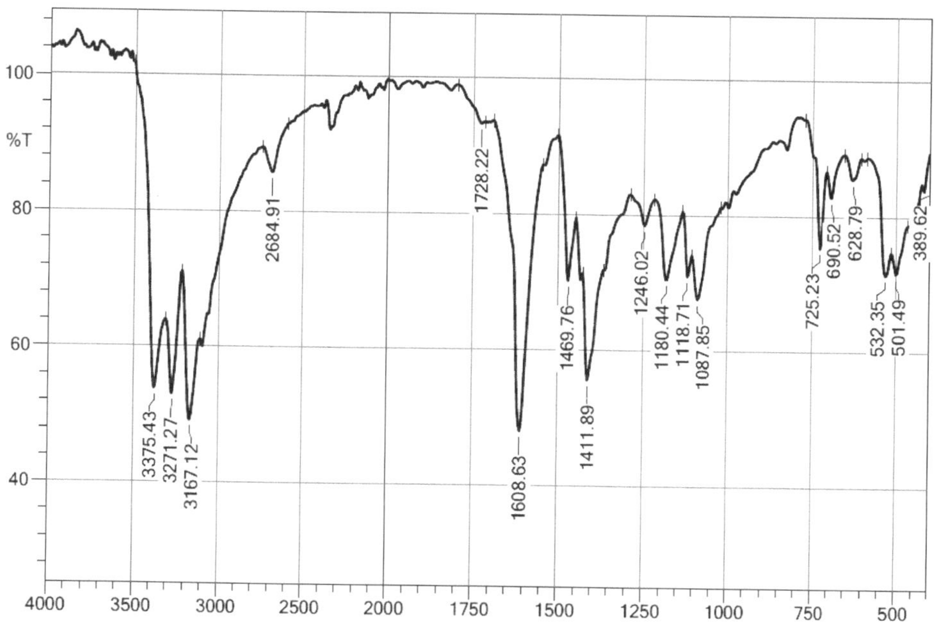

IR spectrum (6d)

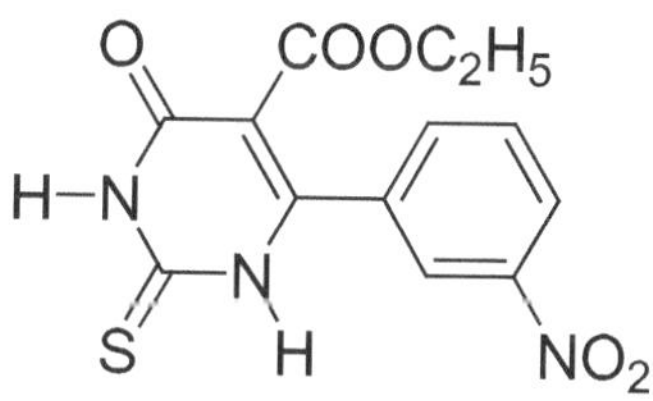

ethyl 1,2,3,4-tetrahydro-6-(3-nitrophenyl)-4-oxo-2-thioxopyrimidine-5-carboxylate (6d)

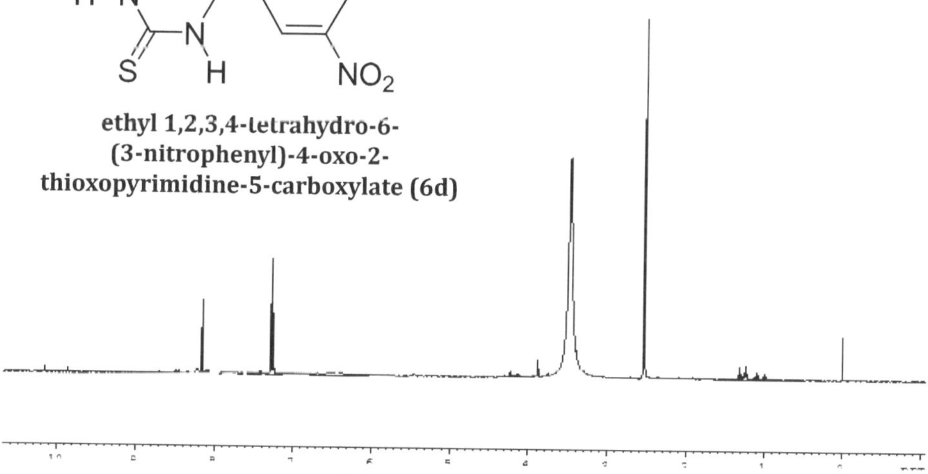

^{1}H-NMR spectrum (6d)

Figure 2.18: Spectrum of Dihydropyrimidine Carboxylate (6d).

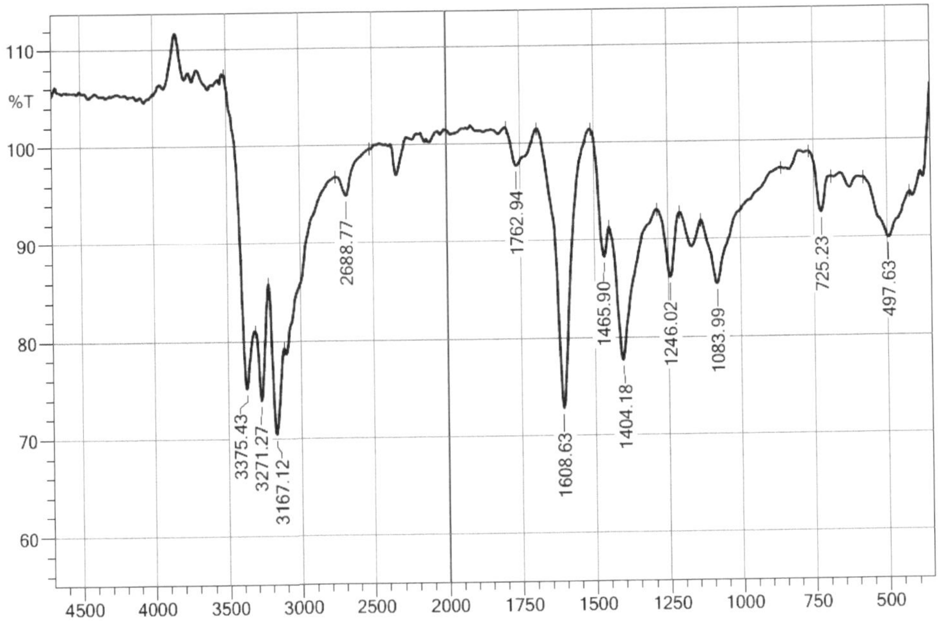

IR spectrum (6e)

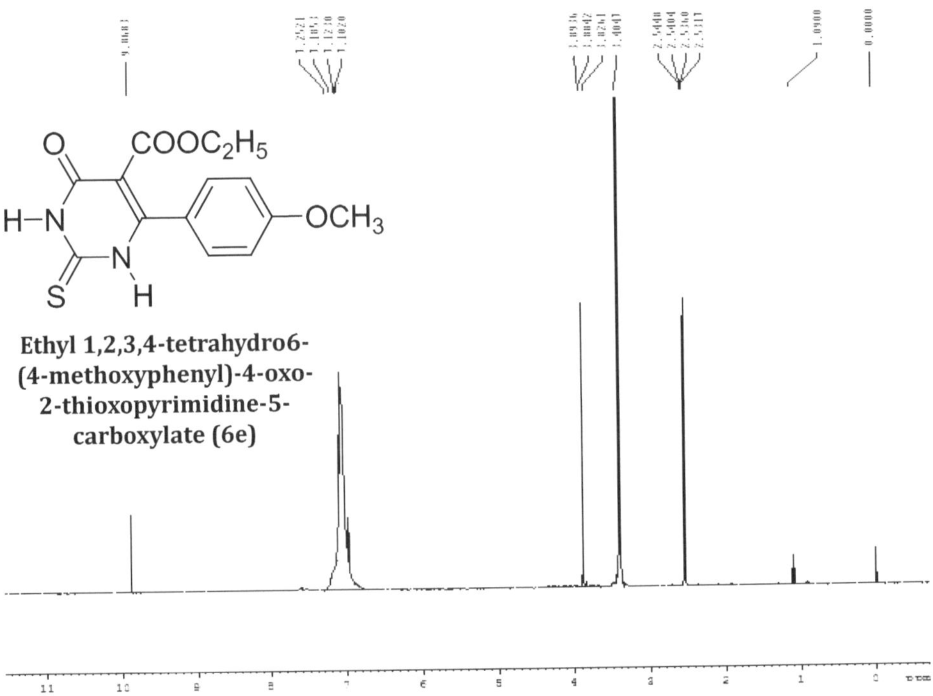

Ethyl 1,2,3,4-tetrahydro6-(4-methoxyphenyl)-4-oxo-2-thioxopyrimidine-5-carboxylate (6e)

^{1}H-NMR spectrum (6e)

Figure 2.18: Spectrum of Dihydropyrimidine Carboxylate (6e).

Ethyl 6-(4-chlorophenyl)-1,2,3,4-tetrahydro-4-oxo-2-thioxopyrimidine-5-carboxylate (6c) : Yield 88 per cent, mp 230-232 °C, **IR (cm^{-1}):** 3375, 3271, 3167, 2689, 1762, 1609, 1465, 1404, 1246, 1083, 729, 629 cm^{-1}. **^{1}H NMR (DMSO-d$_6$ 400 MHz):** 8.22 ∂ (s, -NH), 7.49 ∂ (d, 1 H, ArH), 7.39 ∂ (d, 1 H, ArH), 7.35 ∂ (d, 1 H, ArH), 7.33 ∂ (d, 1 H, ArH), 3.82 ∂ (q, OCH_2), 1.27 ∂ (t, -CH_3).

Ethyl 1,2,3,4-tetrahydro-6-(3-nitrophenyl)-4-oxo-2-thioxopyrimidine-5-carboxylate (6d) : Yield 72 per cent, mp 230-232 °C, **IR (cm^{-1}):** 3375, 3271, 3167, 2685, 1728, 1608, 1470, 1411, 1246, 1180, 1118, 1088 cm^{-1}. **^{1}H NMR (DMSO-d$_6$ 400 MHz):** 8.17 ∂ (s, -NH), 7.88 ∂ (d, 1 H, ArH), 7.86 ∂ (d, 1 H, ArH), 7.84 ∂ (d, 1 H, ArH), 7.82 ∂ (d, 1 H, ArH), 3.88 ∂ (q, OCH_2), 1.31 ∂ (t, -CH_3).

Ethyl 1,2,3,4-tetrahydro-6-(4-methoxyphenyl)-4-oxo-2-thioxopyrimidine-5-carboxylate (6e) : Yield 80 per cent, mp 176-178 °C, **IR (cm^{-1}):** 3375, 3271, 3167, 2689, 1767, 1608, 1466, 1404, 1246, 1084 cm^{-1}. **^{1}H NMR (DMSO-d$_6$ 400 MHz):** 9.86 ∂ (s, -NH), 7.25 ∂ (d, 1 H, ArH), 7.18 ∂ (d, 1 H, ArH), 7.12 ∂ (d, 1 H, ArH), 7.10 ∂ (d, 1 H, ArH), 3.4 ∂ (s, 3H, -OCH_3), 3.49 ∂ (q, -OCH_2), 1.09 ∂ (t, -CH_3).

2.5 Conclusion

In conclusion, substituted dihydropyrimidine carbonitrile (**4a-e**), dihydropyrimidine-2(1H)-one (**5a-e**) and dihydropyrimidine carboxylates (**6a-e**) were synthesized successfully by new protocol one pot effective and systematic protocol by use of simply accessible, low cost research grade chemical reagent purchased from Loba Chemie, Merck, *etc.* The prepared analogues were refined from recrystallization technique, characterized and confirmed by different spectroscopic techniques such as Infrared (IR) and ^{1}H-NMR spectroscopy. Melting point was found to correlate with reported literature values.

2.7 References

1. (a) A. Hantzsch, *Ber.* 14, **1881**, 1637. (b) A. Hantzsch, Justus Liebigs *Ann. Chem.*, 215, **1882**, 1.
2. (a) E. Knoevenagel, *Ber.* 27, **1894**, 2345. (b) E. Knoevenagel, *Justus Liebigs Ann.Chem.*, 281.**1894,** 25. (c) E. Knoevenagel *Chem. Ber.*, 29, **1896**, 172. (d) E. Knoevenagel, *Ber.Dtsch. Chem. Ges.*, 31, **1898**, 2585. (e) E. Knoevenagel, *Ber. Dtsch.Chem. Ges.*, 31, **1898**, 2596.
3. P. Biginelli, *Ber.*, 24, **1891**, 317.
4. G. W. Kenner, A. Todd, Pyrimidine and its derivatives. In Heterocyclic Compounds.,Elderfield, R. C. Ed.; John Wiley and Sons, Inc.: New York, 6, **1957,** 239.
5. L. E. Hinkel, D. H. Hey, *Recl. Trav. Chim.*, 48, **1929**, 1280.
6. (a) P.Biginelli, *Gazz. Chim. Ital.*,23, **1893**, 360. (b) C.O. Kappe, *Tetrahedron.*, **1993**, 49, 6937. (c) C. O. Kappe, *Acc. Chem. Res.*, 33, **2000**, 879. (d) C. O. Kappe, *Eur. J. Med. Chem.*, 35, **2000**, 1043. (e) C.O. Kappe, *QSAR Comb. Sci.*, 22, **2003**,

630.(f) C.O. Kappe, A. Stadler, *Org. React.*, 63, **2004**, 1. (g) D. Dallinger, A. Stadler C.O. Kappe, *Pure Appl. Chem.*, 76, **2004**, 1017.

7. (a) K. Folkers, T.B. Johnson, J. Am. Chem. Soc., 55, **1933**, 3784. (b) D. Dallinger, C.O. Kappe, *Pure Appl. Chem.*, 77, **2005**, 155. (c) I. T. Phucho, A. Nongpiur, S. Tumtin, R. Nongrum, R. L. Nongkhlaw, *Rasayan J. Chem.*, **2009**, 2, 662. (d) J-P. Wan, Y. Liu, *Synthesis.*, 23, **2010**, 3943.
8. F. J. Sweet, D. Fissekis, *J. Am. Chem. Soc.*, 95, **1973**, 8741.
9. B. C. O'Reilly, K. S. Atwal, *Heterocycles.*, 26, **1987**, 1185.
10. K. S. Atwal, B. C. O'Reilly, J. Z. Gougoutas, M. F. Malley, *Heterocycles.*, 26, **1987**, 1189.
11. K. S. Atwal, G. C. Rovnyak, B. C. O'Reilly, J. J. Schwartz, *Org. Chem.*, 1989, 54.
12. C. O. Kappe, *J. Org. Chem.*, 62, **1997**, 7201.
13. V. I. Saloutina, Burgarta, V. Ya. O. G. Kuzuevaa, C.O Kappe, O.N. Chupakhin, *J. Fluor. Chem.*, 103, **2000**, 17.
14. P. B. Chaudhari, S. D. Jadhav, R. P. Dhavale, M. S. Batra, S. S. Shaha, R. B. Ingale, N. M. Bhatia, *Ame-Eur J. Sci Res.*,7(2), **2012**, 69-76.
15. B. A. El Gazzar, H. N. Hafez, *Acta Chem Slov*, 56, **2008**, 359-371.
16. B. S. Holla, K. V. Malini, B. S. Rao, B. K. Sarojini, N. S. Su, *Eur. J. Med. Chem.*, 38, **2003**, 313-318.
17. G. Regnier, R. Canevari, J. Le Douarec, S. Holstrop, J. Daussy, *J. Med. Chem.*, 15, **1972**, 395-401.
18. C. A. Winter, E. A.Risley, G. W. Nuss, *Proc. Soc. Exp. Bio. Med.*, 111, **1962**, 544-547.
19. M. M. Mojtahedi, M. R. Saidi, J. S. Shirzi, M. Bolourtchian, *Syn. Commun.*, 32, **2002**, 851-855.
20. F. Frankenne, H. W. Krell, F. Grams, J. M. Foidart, A. Noel, *Clin. Cancer Res.*, 10, 2004, 4038-4042.
21. M. Huang, Y. Wang, M. Collins, B. S. Mitchell, L. M. Graves, *Mol. Pharmacol.*, 62, **2002**, 463-468.
22. N. Vonbubnoff, D. R. Vesch, W. T. Miller, W. Li, J. Sanger, C. Peschel, W. G. Bornmann, B. Clarkson, J. Duyster, *Cancer Res.*, 63, **2003**, 6395.
23. P. Scheider, S. Hawser, K. Islam, *Biorg. Med. Chem Lett.*, 13, **2003**, 4217.
24. Y. P He, F. Chin, G. Sun, Y. Wang, E. D. Clercq, J. Balzarini, C. Pannecouque, *Biorg. Med. Chem. Lett.*, 13, **2004**, 3173-3176.
25. H. T. A. Mohsen, F. A. Ragab, M. M. Ramala, H. I. El Diwani, *Eur. J. Med. Chem.*, 45, **2010**, 2336-2344.

26. S. A. Galal, A. S. Adbelsamie, H. Tokuda, N. Suzuki, A. Lida, M. M. El Hefnawi, R. A. Ramadan, H. I. El Diwani, *Eur. J. Med. Chem.*, 46, **2011**, 327-340.
27. G. C. Rovnyak, S. D. Kimball, B. Beyer, G. Cucinotta, J. D. Dimarco, J. Gougoutas, A. Hedberg, M. Mulley, J. P. McCarthy, R. Zang, S. Moreland, *J. Med. Chem.*, 38, **1995**, 119-129.
28. A D Patil, N V Kumar, W C Kokke, M F Bean, A J Freyer, C D Brosse, S Mai, A Truneh, D J Faulkner, B Carte, A L Breen, R P.Hertzberg, R K Johnson, J W Westley, B C Potts, *J Org Chem*, 60, **1995**, 1182.
29. T. U. Mayer, T. M. Kapoor, S. J. Haggarty, R. W. King, S. L. Schreibers, T. J. Mitchinson, *Science*, 286, **1999**, 971.
30. M. Nasef, S. J. A. El. Naem, B. A. El Shbrawy, *Egypt J. Pharma. Sci.*, 33-34, **1992**, 463.
31. Habele, R. Riebli, EP 151, **1985**, 130, *Chem. Abst.*, 103, **1985**, 160521.
32. C. Sonnatag, S. Klutchko, J. Shavel, US., 3, **1972**, 649, 644, *Chem. Abst.*, 77, **1977**, 5439.
33. S. Klutchko, J. Shavel, M.V. Strandtmann, US., 3, **1980**, 0660, 644, *Chem. Abst.*, 80, **1974**, 82931.
34. N. Grinev, S. A. Zotava, T. M. Golobova, *Chim Farm Zh*, 19, **1985**, 227, *Chem. Abstr.*184, **1986**, 68700.
35. C.Elliot, V. M. Anthony, D. E. Geroffen, *Chem. Abstr.*, 107, **1987**, 7063.
36. V. Pesteelini, M. Ghelardon, C. Ortolani, E.P. 257, **1986**, 177, *Chem. Abstr.*, 109, **1988**, 65205.
37. Kambe S, Saito K and Kishi H, *Synthesis,* **1979**, 287-289.
38. Patel A, Pasha T Y and Modi A, *International Journal of Pharm Tech Research*, 8, **2015**, 136-143.
39. Murthy Y. L N, Saveri R M R, Parimi A R and Nareesh S, *Org Commun,* 6(1), **2013**, 47-54.
40. Eweas A F, Qasem M A, Abdallah, Emad S I and Hassan, *Journal of Applied Pharmaceutical Science*, 4 (12), **2014**, 102-111.
41. Chitre T S, and Bothara K. G, *Der Chemica Sinica*, 2(2), **2011**, 187-193.
42. Muftah M S and Abuhamrah S E, *International Journal of Advanced Research in Biological Sciences*, 1(9), **2014**, 259-262.
43. Desai S A, Bhatia M S, Choudhari P B and Ingale K B, *International Journal of Pharma Sciences and Research*, 5(6), **2014**, 241-248.
44. H. Yijun, Y. Fengyue, Z. Chengjian, *J. Am. Chem. Soc.*, 127, **2005**, 16386.
45. A. Debache, B. Boumoud, M. Amimour, A. Belfaitah, S. Rhouati, B. Carboni *Tetrahedron letter.*, 47(32), **2006**, 5697.

46. P. V. Anbhule, *Open Catalysis Journal,* 3, **2010**, 83-86.

47. J. C. Bussolari, P. A. McDonnell, *J.Org. Chem.*, 65, **2000**, 6777-6779.

48. S. Patil, S. D. Jadhav, S. Y. Mane, *Inter. J. Org. Chem.*, 1, **2011**, 125-131.

49. B. Gangadasu, P. Narender, B. C. Raju, V. J. Rao, *Indian J. Chemistry*, 45(B), **2006**, 1259-1263.

50. Y. Zhao, Y. Zhao, J. Zhang, H. Liu, L. Liang, *Indian J. Chem.*, 54(B), **2015**, 139-141.

51. N. D. Moirangthem, W. S. Laitonjam, *Indian J. Chem.*, 48(B), **2009**, 1023-1026.

52. M. B. Deshmukh, P. V. Anbhule, S. D. Jadhav, A. R. Mali, S. S. Jagtap, S. A. Deshmukh, *Indian J. Chem.*, 46(B), **2007**, 1545-1548.

53. M. A. Pasha, N. R. Swamy, V. P. Jayashankara, *Indian J. Chem.*, 44(B), **2005**, 823-826.

54. G. Aridoss, Y. T. Jeong, *Bull Korean Chem, Sci.*, 31(4), **2010**, 863-868.

55. M. R. Mohammadizadeh, N. Firoozi, *E-Journal of Chem.*, 8(51), **2011**, 266-270.

56. T. Boumoud, B. Boumound, S. Khovati, A. Belfaitah, A. Desachi, P. Mosset, *Actachim.Slov.*, 55, **2008**, 617-622.

57. A. Shaabani, A. Bazgir, F. Teimouri, *Tetrahedron Letters*, 44, **2003**, 857-859.

58. P. Salehi, M. Dabiri, M. A. Zolfigol, M.A. Bodaghifard, *Tetrahedron Letters*, 44, **2003**, 2889-2891.

59. Garima, V. P. Srivastava, L.Dhar, S. Yadhav, *Tetrahedron Letters*, 51, **2010**, 6436-6438.

60. K. K. Pasunooti, H. Chai, C. N. Jensen, B. K. Gorityala, S. Wang, X. W. Liu, *Tetrahedron Letters*, 52, **2011**, 80-84.

61. S. V. Rashmi, N. C. Sandhya, B. Raghava, M. N. Kumare, K. Mantelingu, K. S. Rangappa, *Syn. Comm.*, 42:3, **2012**, 424-433.

62. Z. L. Shen, X. P. Xu, S. J. Ji, *J. Org. Chem*, 75, **2010**, 1162-1167

63. J. S. Aher, M. R. Gaware, D. D. Lokhande, P. J. Tambade, *J. Basic. Sci*, 4(1), **2017**, 43-46.

Chapter 3

Method, Purification and Measurement of Density and Viscosity

3.1 Introduction of Materials

Different solvents used for our research work such as dimethyl sulphoxide, acetone, toluene *etc.* commercially were purchased from Merck, Qualigens and Loba Chemie, Mumbai and were purified by different methods such as distillation by following reported literature procedure [1].

3.2 Purification of Solvents

Organic solvents purchased commercially contains small quantity of water which is widely present impurity in organic solvent and generally doesn't changes or harms the course of reaction. Similarly traces of other impurities including ethyl alcohol and ether also don't change the condition of desirable chemical process. But if impurities are high and huge volumes of solvents are used then it is more economical and necessary to refine such solvents which are available commercially rather than acquiring high priced AR (Analytical Reagent) grade. Such reagents of appreciable and particular purity can be utilized in refining and isolating processes previously followed by undergoing analysis spectroscopically and in solution chemistry for preparing solutions of different normality or molarity. Hence it is need to purify required solvents by suitable and easy method of purification and drying over appropriate drying agent which have no chemical reaction on the solvents and doesn't alter the properties of solvents. For some solvents special and specific chemical treatment can be given to eliminate contaminations except water. The purified solvent is allowed to dry and is subjected for distillation with specially designed fractionating column by using low pressure which avoids and protect from further decomposition. Addition of crystalline substances with minute pores also known as molecular sieves of appropriate dimensions is subjected in solvents to eliminate traces of water present during storage.

Reagents and chemicals utilized for our research purpose were purchased from Merck, Qualigens and Loba Chemie, Mumbai as analytical and research grade.

3.2.1 Water

Purification Method [1]

Water is an important solvent used in our research work as many of the compounds are easily soluble without undergoing any change in chemical properties of the compounds. Hence water is truly called as "Universal solvent". Owing to this reason purification of water is of prime importance. So we purified water by redistilling in quick fit apparatus over alkaline $KMnO_4$ to oxidise organic impurities followed by further distillation over H_2SO_4. The measured electric conductance of purified and distilled water was found to be in the range 5.53 x 10^{-6} ohm^{-1} cm^{-1} to 5.58 x 10^{-6} ohm^{-1} cm^{-1}.

3.2.2 Dimethyl Sulphoxide (DMSO)

Dimethyl sulphoxide is colourless liquid, highly water soluble, hygroscopic and having typical odour of organosulphur with molecular formula $(CH_3)_2SO$. Melting point and boiling point of dimethyl sulphoxide is 18.5°C and 189°C respectively. It was synthesized by Russian chemist Alexender Zaytsev and reported in 1867 [3].

Importance

Owing to high miscibility of dimethyl sulphoxide in water and for dissolving organic as well as inorganic compounds, preparation of our compounds was done in dimethyl sulphoxide along with spectroscopic techniques. It is clear that DMSO is highly soluble in water in different proportions [4] giving exothermic reaction the physical properties of this mixture are not linear corresponding to each constituents as other solvent-water systems. DMSO properties have been practised for different gases [5], liquids, certain salts [6] and polymeric materials [7-11]. Owing to such indistinguishable properties, DMSO is considered to be of prime interest and is known as super solvents due to its wide applications as solvents in chemical and biological processes including plants, animals and intermediate [12]. Because of S=O group, DMSO is highly polar, aprotic and undergoes self-association through different forces possessing high dipole moment (ε= 46.6) at 25°C and ability to form strong hydrogen bond [13, 14, 15]. DMSO study is of prime importance due to its applications in medicinal field [16] as it easily penetrates into biological membranes, fascinates chemical transport into biological membranes and also has cryoprotective effect on biological systems [17, 18].

Dimethyl sulphoxide is also considered as strong and powerful solvents for different compounds [19, 20] and polymeric materials possessing high molecular weight polymers [21] and often has application in conducting reaction [22]. In pharmaceutical field, DMSO is used due to unique property of skin penetration which boosts absorption of many drugs through skin [23-26]. Normally DMSO is used in formulation of drug delivery process and to carry drugs to different disease target which enhances therapeutic activity as toxicological effect of use of dimethyl sulphoxide is tolerable [27]. DMSO is also used topically in applied medicines on skin for herpes infections [28]. Dimethyl sulphoxide has anti-tuberculosis activity [29]. The study of DMSO is important due to applicability in medicines [30, 31]. DMSO is also utilized in treatment of cancerous cells [32]. Dimethyl sulphoxide has been effectively utilised as topical analgesics, antifungal, anti-inflammatory and antioxidant [33].

Purification Method [1, 2]

Dimethyl sulphoxide (DMSO) available commercially was purchased from Qualigens Pharma Pvt. Ltd, Mumbai and kept in activated alumina (Al_2O_3) for drying and filtration. The resultant filtered dimethyl sulphoxide was distilled by using calcium hydroxide at low pressure and stored store over Type 4A molecular sieve. The contamination level present in DMSO was checked from density and refractive index data which were compared standard literature value as shown in Table 3.4.

3.3 Measurement of Density and Viscosity

3.3.1. Density

Density of any liquid is termed as ratio of mass of liquid to that of volume of liquid and has unit, g. cm^{-3} in CGS system or Kg. m^{-3} in MKS system. It is mostly measured from specific volume of liquid measurement using specially designed bottle also known as pycnometer or Archimedes principle which involves "Sinkers" dipped in liquids whose density is to be estimated. Small changes in densities are determined by measuring rise or fall of liquid in small quartz float. Approximately densities are determined by specially designed hydrometers.

When small amount of liquids is available and high accuracy is required then density of liquid is excellently determined by pycnometer having variable shapes and sizes. This pycnometer apparatus was first invented by Spengel and then modified by Ostwald. Perkin, Bonsfield are different pycnometer forms. Many pycnometer used at different temperatures are single capillary. Here in our present study we are using bicapillary pycnometer for measuring the densities of liquids and solutions

3.3.2 Calibration Method

Calibration is done to accurately measure the exact volume of new measuring vessel. The capacity of such vessels changes with respect to temperature. Hence in case if a measuring vessel is not calibrated at given temperature then an positive or negative error may be observed. The calibration of measuring vessel can be achieved by different methods as reported in literature [34, 35, 36].

A bi-capillary used for our research work comprising of bulb having approximately 10-11 ml capacity and capillary having internal diameter 1mm was selected. The schematic picture of bicapillary pycnometer is shown in Figure 3.1. Before calibration, pycnometer was thoroughly washed with solution of chromic acid prepared freshly to eliminate any impurity and obstacles in the capillary that may enable a steady flow of liquid and solution and also helps in without sticking of drop behind followed by washing with triple refined water and acetone. By using stream of warm air generated from electrical hair blower, pycnometer was dried and kept in electrical chamber maintained at 60-80°C for nearly half an hour. The pycnometer was cooled and empty weight was determined by using calibrated electronic balance having sensitivity of 0.001g. After weighing pycnometer was slowly filled with triple refined water by immersing the end of limb "A" in beaker containing triple refined water. Excess amount of water sticked to glass of pycnometer was wiped by using specially designed wipping filter paper. Precaution was taken that no particles of paper were stricken on the surface of pycnometer r. Again the pycnometer was weighed and the weight of water was calculated by taking the difference of weight of filled pycnometer and weight of empty pycnometer. Now the pycnometer containing triple refined water kept vertically in temperature controlled water bath as shown in Figure 3.2 to attain thermal equilibrium. The height of water in each arm was noted by using travelling microscope as 'h_1' and 'h_2' in cm. The overall height 'h' cm was calculated by taking summation of height of water in each arm of pycnometer as $h = h_1+h_2$ for every degree rise in temperature from 298.15 to 313.15 K. Weight and standard density of super refined water at temperatures ranging from 298.15 to 313.15 K, gives volume of water at corresponding temperature from formula $v = w/\rho$. The graph of total height (x-axis) against calculated volume (y-axis) gives value of slope 0.0059 and intercept as 10.93 with correlation coefficient 0.99986.

Now the pycnometer was filled with liquids like toluene and acetone whose density is to be determined and the procedure was repeated for these liquids. From the values of slope and intercept, volume of these liquids was calculated by using formula, volume (V) = (slope x total height) + intercept at 298.15, 303.15, 308.15 K. The density of acetone and toluene was calculated by using formula $\rho = w/v$ where "w"= weight of liquid, "v"= volume of liquid.

Table 3.1: Literature and Observed Density of Acetone and Toluene

Sl.No.	*Temp. K*	*Acetone g. cm^{-3}*		*Toluene g. cm^{-3}*	
		Lit.	*Observed*	*Lit.*	*Observed*
1	298.15	0.7870[1]	0.78742	0.8623[1]	0.86242
2	303.15	0.7790[1]	0.77864	0.8574[1]	0.85752
3	308.15	0.7729[1]	0.77259	0.85271	0.85281

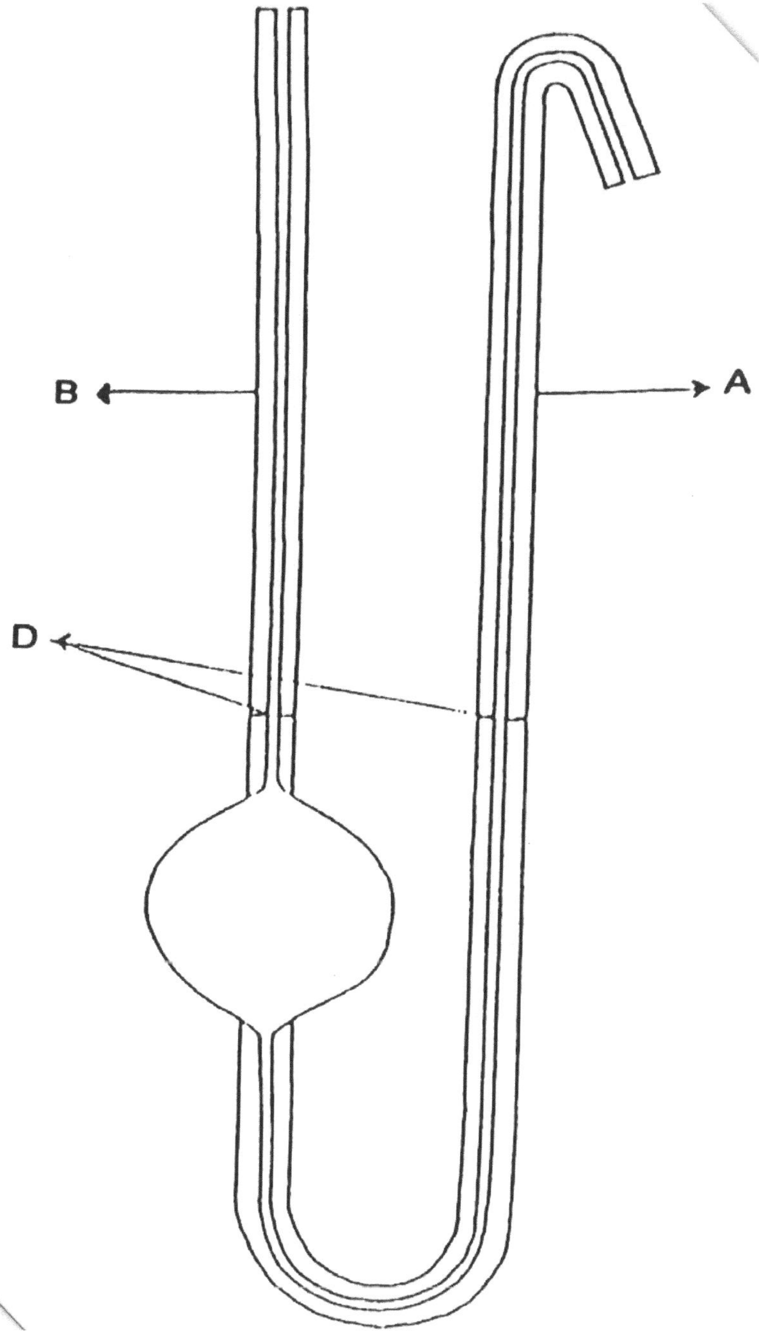

Figure 3.1: Bicapillary Pycnometer.

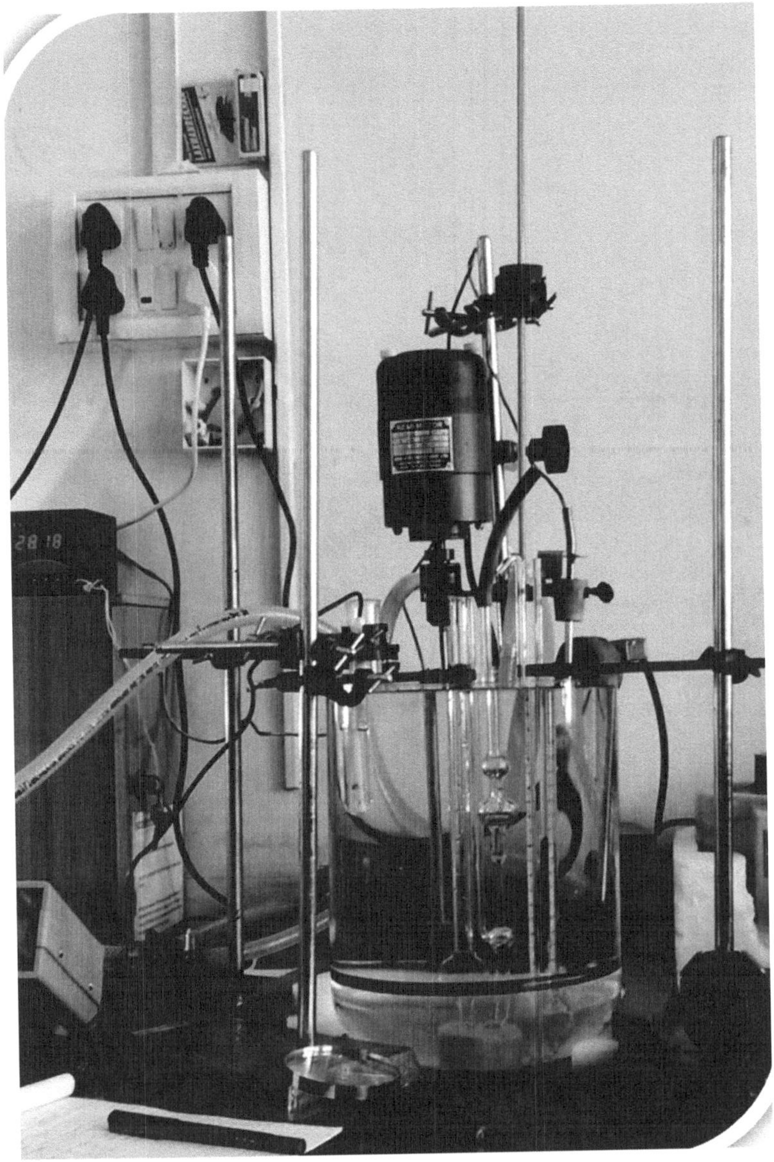

Figure 3.2: Experimental Setup.

It is found that this density values have good agreement with reported values as tabularised in Table 3.1 with error of ± 1 x 10^{-4} g.cm^{-3}.

3.3.3 Density Measurement of Solutions

Now washed, cleaned and free from water pycnometer was filled with experimental solution of different concentration by the method as described above.

This pycnometer is kept in thermally controlled water bath for 20-30 minutes for equilibrium. The overall height "h" cm of solution is calculated as $h = h_1 + h_2$ over entire range of temperature 298.15, 303.15, 308.15, 313.15 with the help of travelling microscope as described above. By knowing overall height "h" from pycnometer, volume was calculated using formula v = (slope x height) + intercept, at different temperature and analogous density are calculated using $\rho = w/v$. The same procedure was repeated for determination of densitometric measurement of other solutions at divergent temperatures.

3.3.4 Viscosity

Viscosity is crucial in technical field and is defined as resistance offered by liquid to flow. It is inversely proportional to fluidity of liquid. Viscosity is measured by using different viscometers such as Ostwald viscometer, Ubbelohde viscometer. Two different methods are adopted for viscometric study of liquids

1. **Capillary Method**

In capillary method, the fluid is allowed to flow through narrow tube due to hydrostatic or applied pressure. Poiseuille law relates the rate of flow through capillary to the viscosity of the liquid is the basis for capillary method and is given by equation as shown below:

$$v = (\pi P r^4 t)/(8 \eta l)$$

where,

P = Pressure.

l = Length of capillary.

η = Coefficient of viscosity.

r = is the radius of the capillary.

2. **Falling Body Method using Spherical Ball**

In this method a spherical ball is allowed to fall freely through measured distance of viscous medium liquid and its velocity is determined. The viscous drag resulting due to falling liquid results in formation of restoring force and is determined by Stroke's law as shown below:

$$\eta = 2\, g\, r_s^2\, (\rho_2 - \rho_1)/9\, U_t$$

where,

r_s = Radius of sphere

U_t = Terminal velocity of falling sphere

ρ_1 = Density of viscous liquid

ρ_2 = Density of sphere

g = Acceleration due to gravity

For our present work, viscosity of solutions of different concentration was achieved by capillary flow method at four different temperature using Ubbelohde viscometer [37, 38] also known as suspended – level viscometer. The schematic picture of suspended level Ubbelohde viscometer is shown in Figure 3.3. It consist of "U"-shaped glass tube with reservoir and measuring bulb with capillary on two different sides. Flowing time of viscous liquid between "A" and "B" marks under gravity is evaluated. This type of viscometer has third arm "H" also known as pressure equalization arm is present which rely on height and not on volume taken in viscometer. This viscometer is called "suspended level viscometer because the capillary is suspended above the reservoir. Viscometric constant values are same at all temperatures which is an important property during evaluation of viscosity. It is also applicable to measure dynamic viscosity of less opaque liquids in which meniscus is visible [34]. Another advantage of using this viscometer is speed, accuracy with small sample size, less error deviation and cheaply available.

3.3.5 Calibration Method

Viscometer used for viscosity measurement comprises of bulb with 20 ml capacity along with capillary of about 80-95 mm long having internal diameter of 0.5 mm. Before calibrating the viscometer, it is properly washed with freshly prepared chromic acid solution to remove any impurities leading to obstruct for the smooth flow in capillary without sticking drops behind the surface of viscometer followed by washing with refined water and by acetone. Electrically operated hot hair blower is used to remove traces of water from viscometer. This dried viscometer is kept in hot oven regulated and maintained at 70-80°C temperature for 30 minutes.

The dried viscometer is filled with triple refined water though vertically fixed tube "F" and kept in thermally controlled water bath for 10-20 minutes for thermal stability. The water was drawn above the mark "A" of capillary tube by holding finger on the upper portion of tube "H". Now the figure was slowly adjusted and removed so that water flows steadily down the past mark "A" followed by measuring time between "A" to "B" mark on the side of bulb "D" at 298.15, 303.15, 308.15 and 313.15 K. By using efflux time and standard densities, viscosities of water at various temperature was calculated using following equation

$$\text{Viscosity } (\pi) = \rho\ (at\text{-}b/t)$$

where,

$$\frac{\eta * t}{\rho} = a\,t^2 - b$$

"a" and "b" are viscometric constant.

"t" is the efflux time.

"ρ" is the density.

Graph of $\eta * t/\rho$ against t^2 was obtained gives the value of viscometer constants "a" as slope of 0.0051 and "b" as intercept of -4.5152 respectively with correlation

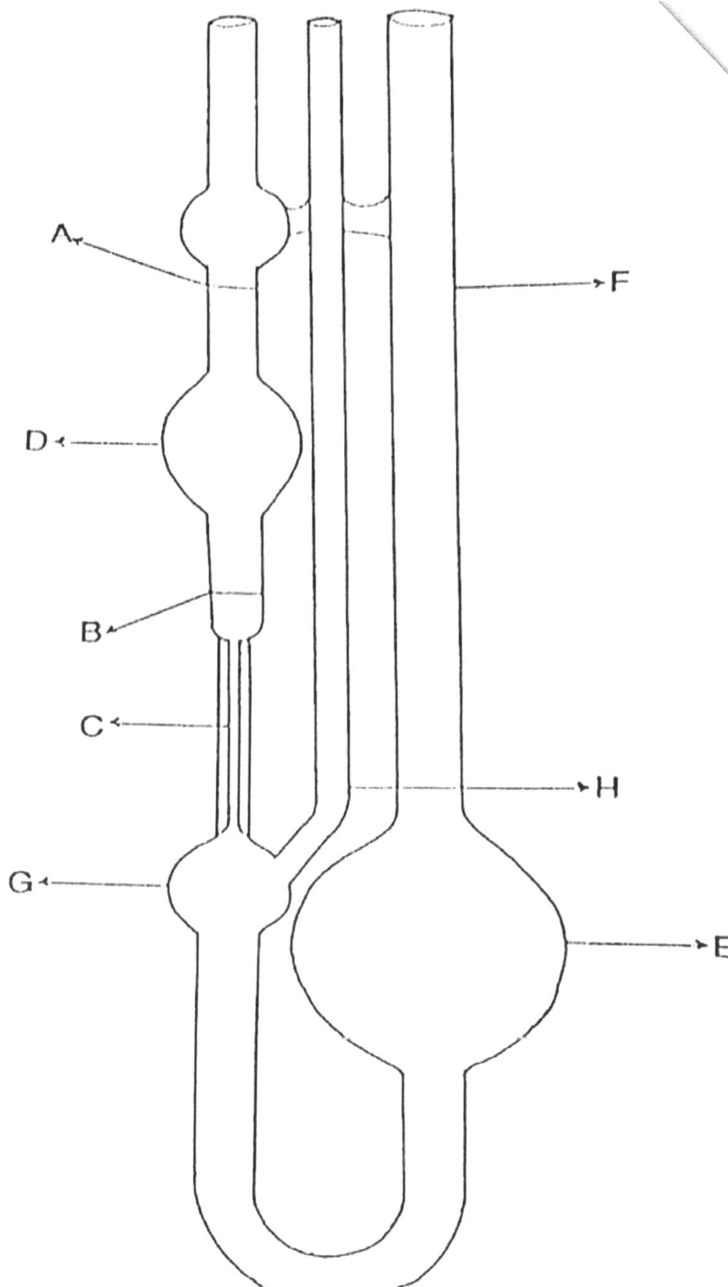

Figure 3.3: Ubbelohde Viscometer.

coefficient of 0.999948. Viscosity of purified acetone and toluene were measured in similar manner by using the equation as stated above. We found that the viscosity values determined for these liquid were close to the reported literature values at 298.15 to 313.15 K temperatures as shown in Table 3.2.

Table 3.2: Literature and Observed Viscosity of Acetone and Toluene

Sl.No.	Temperature K	Viscosity (η) of Acetone (cP)		Viscosity (η) of Toluene (cP)	
		Lit.	Observed	Lit.	Observed
1	298.15	0.306[1]	0.30564	0.560[1]	0.55906
2	303.15	0.295[1]	0.29369	0.530[1]	0.52735
3	308.15	-	0.27870	0.500[1]	0.50074

3.3.6 Viscosity Measurement of Solutions

Properly wash was given to viscometer by using chromic acid solution to eliminate any obstacles in the capillary that restricts steady flow of liquid followed by washing with purified water and acetone. Drying was performed by using electrically operated hair blowing machine and subjected in oven at about 70-75°C for 30 minutes. Solution under investigation was poured though vertically fixed tube "F" in and kept in thermally controlled water bath nearly for 15-20 min for thermal stability. The test solution was drawn above the mark "A" of capillary tube by holding finger on the upper portion of tube "H". Now the figure was slowly adjusted and removed so that solution flows steadily down the past mark "A" followed by measuring time between "A" and "B" mark on the side of bulb "D" at 298.15, 303.15, 308.15 and 313.15 K.

The viscosity of test solutions at various temperatures was estimated using the equation as shown above. The time was measured three times and average value was used in calculations.

3.4 Preparation of Solutions

3.4.1 Preparation of Aqueous Dimethyl Sulphoxide

Purified DMSO was added to triple refined water to produce 60 per cent aqueous DMSO solution (60 per cent dimethyl sulphoxide and 40 per cent water v/v). This prepared aqueous solution was stored carefully neglecting the effect of humidity. The densitometric and viscometric studies were performed using pycnometer and Ubbelohde viscometer respectively at 298.15, 303.15, 308.15 and 313.15 K temperature. The densitometric and viscometric values of 60 per cent aqueous DMSO were determined and are shown in Table 3.5.

3.4.2 Preparation of Stock Solution of Dihydropyrimidine and its Derivatives

Five different substituted dihydropyrimidine carbonitriles, dihydropyrimidin-2(1H)-ones and dihydropyrimidine carboxylates were properly weighed and solutions in 60 per cent aqueous dimethyl sulphoxide were prepared having 0.02 M concentration as standard solution.

Table 3.3: Density and Viscosity of water at 298.15 to 313.15 K.

Temp K	*Density (ρ_o) g cm^{-3}*		*Viscosity (η_o)cP*	
	Expt.	*Lit*	*Expt.*	*Lit.[a]*
298.15	0.997051	0.99705 [a] 0.997051 [d]	0.895	0.891
303.15	0.995655	0.99565 [a] 0.995678 [d]	0.805	0.798
308.15	0.994037	0.99403 [a] 0.994064 [d]	0.725	0.719
313.15	0.992236	0.99222 [a] 0.992247 [d]	0.655	0.653

Table 3.4: Density and Viscosity of Dimethyl Sulphoxide at 298.15 to 313.15 K

Temp K	*Density (ρ_o) g cm^{-3}*		*Viscosity (η_o)cP*	
	Expt.	*Lit.*	*Expt.*	*Lit [a]*
298.15	1.09545	1.0955 [a] 1.09562 [g]	1.99089	1.991
303.15	1.09018	1.0904 [a] 1.0906 [f]	1.80486	1.806
308.15	1.08531	1.0854 [a] 1.08548 [e]	1.69330	1.690
313.15	1.08037	1.0803[a]	1.51037	1.511

Table 3.5: Density and Viscosity of 60 per cent Aqueous Dimethyl Sulphoxide Binary Mixture at 298.15 to 313.15 K

Temp K	*Density (ρ_o) g cm^{-3}*		*Viscosity (η_o)cP*	
	Expt.	*Lit.*	*Expt.*	*Lit [b]*
298.15	1.082461	1.08231 [b, h, i]	3.37059	3.42489
303.15	1.08135	1.08111 [b]	2.99112	2.9921
308.15	1.07298	1.0728 [b]	2.58360	2.6712
313.15	1.06547	1.06521 [b]	2.24357	2.3483

Ref.: (a = 43, b = 44, c = 45, d = 46, e = 47, f = 48, g = 49, h = 50, i = 51).

Concentrations from 0.002 M to 0.01 M of each five substituted dihydropyrimidine carbonitriles, dihydropyrimidin-2(1H)-ones and dihydropyrimidine carboxylates were prepared from standard solution by using $M_1V_1 = M_2V_2$. By preparing standard solution of particular substituted dihydropyrimidine carbonitriles, dihydropyrimidin-2(1H)-ones and dihydropyrimidine carboxylates on dilution give required concentration and measurement was done on the very same day to avoid degradation and any type of error.

3.5 References

1. Vogel's *Textbook of practical Organic Chemistry*, 5th Ed.
2. R.G. LeBel and D.A.I. Goring, *J. Chem. Eng. Data*. 7 (1), **1962**, 100-101.
3. Kathrin-Maria Roy "Sulfones and Sulfoxides" in *Ullmann's Encyclopedia of Industrial Chemistry*, **2002**, Wiley-VCH, Weinheim.
4. I. Mellan, "*Industrial solvents," 2nd ed., Reinhold*, New York, **1950**.
5. T. H. Smedslund, *Finska Kemistsamfundets Medd*. 59, **1950,** 40.
6. F. A. Abadie-Maumert, *Papeterie* 81, **1959**, 187.
7. H. W. Giertz, J. MacPherson, Norsk Skogsin, 10, **1956**, 348.
8. E. Hagglund, B. Lindberg, J. MacPherson, *Acta Chem. Scand*. 10, **1956,** 1160.
9. S. Hossain, *Pulp Paper Mag. Can*. 59, **1958**, 127.
10. P. J. Killion, J. F. Foster, *J. Polymer Sci*., 40, **1960**, 65.
11. R. St. J. Manley, *Svensk Papperstidn*, 61, **1958**, 96.
12. G. A. Krestov, *Thermodyn. of solutions, Ellis Harwood Limited*, England, **1991**.
13. B. V. K. Naidu, K. C. Rao, M. C. S. Subha, *J. Chem. Eng. Data*. 47, **2002**, 379.
14. J. B. Kinsinger, M. M. Tannahill, M. S. Greenberg, A. I. Popov, *J. Phys. Chem*. 77, **1973**, 2444-2449.
15. Y. Marcus, *Introduction to liquid state Chemistry, Wiley-Interscience*, New York, **1977**.
16. H. H. Szmant, S. W. Jacob, E. E. Rosenbaum, D. C. Wood (Eds.), *Dimethyl sulphoxide, Marce Dekker*, New York, NY, **1971**, 1-98.
17. J. T. Lal, F. W. Lau, D. Robb, P. Westh, G. Nielsen, C. Tranudum, A. Hvidt, Y. Koga, *J. Solution Chem*. 24, **1995**, 271-277.
18. R. C. Scaduto, *Free Radical Med*. 18 (2), **1995**, 271-277.
19. H. L. Schaefer and Schaffernicht, *Angew. Chem*. 72, **1960**, 618.
20. A. J. Parker, *Quart. Rev*. (*London*), 16, **1962**, 163.
21. D. Martin, A Weise and H. J. Nicolas, *Angew. Chem. Intern. Ed. Engl*. 6, **1967**, 318.

22. W. S. MacGregor, Ann. N. Y. *Acad. Sci.,* 141, **1967**, 3-12.
23. S. W. Jacob, E. E. Rosenbaum, Eds. Ann. N. Y. *Acad. Sci.* **1976**, 141.
24. S. W. Jacob, R. Herschelr, Eds. Ann. N. Y, *Acad, Sci.* **1975**, 243.
25. De La Torre, J. C., Eds. Ann. N. Y. *Acad. Sci.* **1983**, 411.
26. S. W. Jacob, R. J. Herschelr and M. G. Bischel, Belg. Pat. 644, **1964** to Crown Zellerbach Corp.; through *Chem. Abstr.*, 63, **1965**, 9759.
27. J. L. Shen, Z. F. Li, B. H. Wang, Y. M. Zhang, *J. Chem. Thermodyn.*, 32, **2000**, 805-819.
28. J. S. Aguilar, D. Roy, P. Ghazal and E. K. Wagner, *BMC Infect. Dis.* 2, **2002**, 9-19.
29. C. Jagannath, V. M. Reddy, P. R. Gangadharan, *J. Antimicrob. Chemother.* 35, **1995**, 390-391.
30. S. W. Jacob, E. E. Rosenbaum, and D. C. Wood, *Dimethyl sulphoxide, New York: Marcel Dukker* **1971**.
31. J. T. Lal *et al., J. Sol. Chem.*, 24 (1), **1995,** 89-102.
32. R. C. Scaduto, *Free radical Biol. Med.,* 18 (2), **1995,** 271-277.
33. *D. E. Pegg, J.G. Day and G.N. Stacey, Hum. Press. 368,* ***2007****, 39–57.*
34. A. M. James and F. E. Prichard, *"Practical Physical Chemistry"* Longman 3rd ed., **1974**, 19-26.
35. J. Basset, R. C. Denney, G. H. Jeffery and J. Mendham, *"Vogel's Textbook of Quantitative Inorganic analysis",* Longman, 4th ed., **1978**.
36. R. B. Bird, W. E. Stewart and E. N. Lightfoot, *"Transport Phenomena", John Willey and Son. Inc.*, New York, **1960**, 780.
37. M. Z. Kaminsky *Physik Chem*, 5, **1955,** 154.
38. P. S. Nikam, L. N. Shirsat, M. Hasan, *J. Chem. Eng. Data* 43, **1998**, 732-737.
39. J. M. G. Cowie and P. M. Toporowski, *Can. J. Chem.* Vol. 39, **1961**.
40. R. J. Fort and W. R. Moore, *Trans. Faraday Soc.,* 62, **1966,** 1112.
41. J. J. Lindberg and R. Lauren, *Finska Kemistsamfundets Medd.,* 71, **1962**, 37.
42. A. Sacco, E. Mattioli, *J. Solution Chem.* 26, **1997,** 527-535.
43. Lange's Handbook of Chemistry, Sixteenth Edition, James G. Speight, Mc-Graw Hill.
44. T. Zamir and A. Khan, *J. Chem. Soc. Pak*, 27, **2005** 130-136
45. R. B. Torres, A. C. M. Marchiore and P. L. O. Volpe, *J. Chem. Thermodyn.* 38, **2006,** 526-541.
46. L. Korson, W. Drost-Hansen, and F. J. Millero, *J. Phys. Chem.* 75, **1969,** 2016.
47. B. V. K. Naidu, K. C Rao, M. C. S. Subha, *J. Chem. Eng. Data.* 47, **2002**, 379-382.

48. M. A. Saleh, S. Akhtar, M. S. Ahmed, M. H. Uddin, *J. Mol. Liq.* 40, **2002**, 621-635.

49. H. Wang. Liu and J. Huang, *J. Chem. Thermodyn.*, 36, **2004,** 743-752.

50. J. M. G. Cowie and P. M. Toporowski. *Can. J. Chem.*39 **1961**.

51. Maria del Carmen Grande et al, *J. Chem. Thermodynamics*, 39, **2007**, 1049-1056.

Chapter 4

Densitometric and Viscometric Studies of Dihydropyrimidines and its Derivatives

4.1 Introduction

Theoretical models are used to predict the behaviour of liquids [1]. Appropriate measurement and predictions of physical parameters of solutions are very essential in engineering, biological processes and chemical industries [2, 3, 4]. Densitometric, volumetric and viscometric analysis of ideal as well as non ideal solutions are considered to be crucial and major key aspects in solution chemistry.

The apparent and partial molar volume gives an idea about electrolytes at infinite dilute and can be explored to study different interactions such as solute-solute, solute-solvent and solvent-solvent interactions. Investigation of such interactions provides information regarding nature of solvent and solute *i.e.* solute changes or modifies or disturbs the properties of solvent. Therefore for more diluted solutions, apparent molar volume and viscosity are of prime importance to know the environment of intermolecular interactions which gives relation between structure disintegrating and structure making of medicines.

Observed densitometric changes of solutions due to concentration in terms of apparent molar volume $\varnothing_v$ was firstly introduced by Merignac in 1871 [12] as

$$\varnothing_v = (\upsilon - n_1 \upsilon_1^{-0})/n_2 \qquad (5.1)$$

where,

v is the volume of the solution.

$n_1 \upsilon_1^{-0}$ is the volume of water in solution.

n_1 is the number of moles of water.

υ_1^{-0} is the molar volume of water

n_2 is the number of moles of solute in solution.

It is observed, $Ø_v$ increases with increase in temperature along with concentration. The apparent molar volume ($Ø_v$) is determined from molar concentration "C" and densities. For v = 1000 cm³ of solution, "n_2"= number of solute and equals to concentration (C). Hence apparent molar volume ($Ø_v$) of the solution can be calculated from the density data using following relation [13, 14]

$$Øv = \frac{M}{\rho^o} - \frac{1000(\rho - \rho^o)}{C\rho^o} \quad 5.2$$

or

$$Øv = \frac{M}{\rho^o} + \frac{1000(\rho^o - \rho)}{C\rho^o} \quad 5.3$$

where,

"M" = molecular weight of solute.

"C" = molar concentration of solution.

"ρ" = density of solution.

"ρ^0" = density of pure binary solvent.

Molar solution can also be converted into molal concentration by using standard expression; C = 1000 ρm/(1000 + mM) and substituting this value of molar concentration "C" in above equation 5.2 and 5.3, apparent molar volume for molal solution can be calculated by the following formula [15, 16, 17, 18]

$$Øv = \frac{M}{\rho} + \frac{1000(\rho^o - \rho)}{m\rho\rho^o} \quad 5.4$$

$$Øv = \frac{M}{\rho} - \frac{1000(\rho - \rho^o)}{m\rho\rho^o} \quad 5.5$$

where "m" = molality (mol kg⁻¹) of solution. The graph of the apparent molar volume of solute at infinite dilution and concentration dependence expression of the apparent molar volume is shown by Masson equation [19], Redlich-Meyer equation [20] and Owen-Brinkley equation [21].

In 1929, Masson found a valuable empirical (5.6) on the change of $Ø_v$ with square root of molar concentration of dilute solution as

$$Øv = Øv^0 + C^{1/2} \quad 5.6$$

where,

"$Ø_v^0$" = Apparent molar volume at infinite dilution (limiting apparent molar volume) and shows solute-solvent interactions in solution.

"S_v" = Experimental or limiting slope and explains solute-solute interactions in the solution.

The above equation is applicable to concentrated solutions also.

Scott [22] and Geffeken [23] overviewed Øv of solutes using Masson's equation and found that it represents the concentration dependence of the Øv of solutes over a wide range of concentration and temperature.

The concentration dependence of viscosity of dilute solution is given by Jones-Dole [24] semi empirical equation 5.7 as

$$\frac{\eta}{\eta_0} = 1 + AC^{\frac{1}{2}} + BC \quad 5.7$$

$$\eta_r = \frac{\eta}{\eta_0} = 1 + AC^{\frac{1}{2}} + BC \quad 5.8$$

$$\eta_r - 1 = AC^{\frac{1}{2}} + BC$$

$$\frac{\eta_r - 1}{\sqrt{C}} = A + B\sqrt{C} \quad 5.9$$

where,

"η" and "η^0" = viscosities of solution and solvent respectively.

"η_r" = relative viscosity.

"C" – molar concentration of solute.

"A" and "B" = viscosity constant.

From the linear graph of $\eta_r - 1/\sqrt{C}$, the viscosity coefficient "A" (Falkenhagen coefficient) is obtained as intercept while "B" (Jone-Dole coefficient) as slope. The viscosity coefficient "A" coefficient gives solute-solute interaction while "B" coefficient gives contribution arising due to size and molar volume of the solvent along with solute-solvent and solvent-solvent interactions in solution. Hence viscosity "A" and "B" gives effect arises due to solute in solvent.

4.2 Literature Survey

J. Wu and co-workers [25] have studied the density and viscosity of saturated liquid dimethoxymethane were measured over temperature range from (218.15 to 383.15) K with a vibrating-tube densimeter and a calibrated Ubbelohde-type capillary viscometer, respectively. The density results were fitted using the polynomial with an absolute average deviation of 0.13 per cent and a maximum deviation of 0.22 per cent from the correlated equation and viscosity results were correlated as a function of temperature.

R. M. Diguilio *et al.* [26] have reported the densities and viscosities of series of ethanolamines upto normal boiling point of liquid at atmospheric pressure and correlated the data using modified rough hard sphere model for viscosity.

Y. Xu and co- workers [27] studied density and viscosity for binary mixture of *n*-butylammonium acetate ionic liquid (N4AC) with methanol, ethanol and n-butanol at 293.15 to 313.15 under reduced atmospheric pressure. The excess molar volume, viscosity deviations obtained from the experimental obtained data were fitted with Redlich-Kister equation. They found correlation results was in good agreement with the experimental data.

I. S. Khattab [28] *et al.,* have studied density, viscosity, and surface tension of binary mixture of water + ethanol at 293, 298, 303, 308, 313, 318, and 323 K. The molar volume of water+ ethanol mixtures were calculated using measured density values. They used Jouyban-Acree model for mathematical correlation of the data. The relative deviation (RD) was used as an error criterion for density, viscosity, surface tension, and molar volume data. They found that the Jouyban-Acree model correlated physicochemical properties of the mixtures of solvents at different temperatures with acceptable error in calculation.

P. J. Carvalho *et al.* [29] studied experimental data for density, viscosity, refractive index and sound speed of glycols and glymes from 283.15 to 373.15 K at atmospheric pressure. They found that effects like the increase in the number of ethoxy group's increases density, molar volume, viscosity, and refractive index, or the loss of the hydroxyl groups and by substitution of the hydroxyl group's hydrogen with a methyl or ethyl group, lead to a significant decrease in the density, viscosity, and sound speed.

B. Gonzalez and co-workers [30] studied density and dynamic viscosity for different liquids such as methanol or ethanol with water, ethyl acetate and methyl acetate at different temperature ranging from 293.15 to 303.15 K and 0.1 MPa pressure along with the properties of pure component. They calculated excess molar volume, viscosity deviations, free energy of activation for the binary systems and fitted to Redlich-Kister equation to determine root mean square deviations. They used UNIQUAC equation to correlate the experimental viscosity data and UNIFAC-VISCO and ASOG-VISCO method to predict dynamic viscosity of binary mixtures.

M. H. Rausch [31] studied density, liquid kinematic viscosity, and surface tension of the segregated hydrofluoroethers (HFEs) in dependence on temperature from (273.15 to 363.15) K under saturation conditions. They obtained kinematic viscosity of the liquid phase and surface tension on the basis of surface light scattering (SLS) measurements for temperatures between (273.15 and 373.15) K. they discussed effects of the molecular structures of the studied HFEs on the thermophysical data.

E. Vogel [32] generated advanced formula for viscosity in propane, using the reference equation of state for its thermodynamic properties.

A. Hartono and H. F. Svendsen [33] measured densities and viscosities of aqueous diethylenetriamine (DETA) solutions for the entire concentration range and for the temperature range between 293.15 to 363.15 K. They determined excess molar volume from experimental data whereas the excess Gibbs free energy and the excess entropy of flow were determined from the experimental viscosity data by implementing the theory of rate processes of Eyring. They applied Redlich–Kister equation to correlate the excess properties such as excess molar volume and excess excess Gibbs free energy as function of mole fraction of DETA and temperature. They found that it agree very well with the experimental data a higher than amines.

U. Dumanska and M. Krolikowska [34] determined densities and viscosities for binary mixtures of the ionic liquids 1-butyl-3-methylimidazolium thiocyanate, 1-butyl-4-methylpyridinium thiocyanate, 1-butyl-1-methylpyrrolidinium thiocyanate and 1-butyl-1-methylpiperidinium thiocyanate with water from 298.15 to 348.15 K temperature and ambient pressure. They correlated density and viscosity using an emperical second order polynomial and Vogel-Fulcher-Tammann equation respectively. The excess molar volumes and the deviations in viscosity were calculated from experimental values and were correlated by Redlich-Kister polynomial expansions. They also discussed variation in parameters such as composition of mixtures, temperature in terms of molecular interactions along with excess partial molar volumes were calculated. The isobaric expansivities were determined. They found that interactions of thiocyanate with water is not strong as with alcohol which is shown by slightly positive or negative excess molar volumes in these binary systems

R. J, Correia and J. Kestin [35] presented new experimental measurements of the viscosity of aqueous Na_2CO_3 and K_2CO_3 solution in the range of 20-90°C and in the pressure range of 0-30 MPa. They measured effect of pressure on the density of these solutions. They correlated density and viscosity of Na_2CO_3 and K_2CO_3 in terms of temperature, pressure, and concentration.

M. Singh and *et al.* [36] determined density (ρ), apparent molar volume (V_δ), and viscosity (η) of 0.0010 to 0.0018 per cent (*w*/*v*) of bovine serum albumin (BSA), egg albumin, and lysozyme in 0.0002, 0.0004, and 0.0008M aqueous RbI and CsI, and (dodecyl) (trimethyl) ammonium bromide (DTAB) solutions They found that with salt concentration, the density of proteins decreases, and the order of the effect of additives on density is observed as CsI > RbI > DTAB. The trend of apparent molar volume of proteins is found as BSA > egg-albumin > lysozyme for three additives. In general, η values of BSA remain higher for all compositions of RbI than that of egg-albumin for CsI and DTAB. They also found that these orders of the data indicate the strength of intermolecular forces between proteins and salts, which helps for understanding the denaturation of proteins.

Y. P. Hseih [37] *et al.,* measured density and viscosity of the aqueous deep eutectic solvents (DES) systems and determined the parameters of a Vogel–Tamman–Fulcher-type equation and correlated dependence on temperature and composition

S. Han [38] and co-workers studied the density, viscosity and refractive index of sodium lactobionate aqueous solution with respect to molality and temperature. They found that all the above increases when temperature decreases. They used Vogel–Tamman–Fulcher equation to correlate the dependence of density and viscosity on molality and temperature and refractive index. The calculated values showed good agreement with the experimental data for density, viscosity, and refractive index. They determined different volumetric properties of sodium lactobionate mixtures. The relative viscosity and B-coefficient were calculated. They studied the molecule interactions and solution properties.

T. A. Graber and co-worker [39] presented data on the density, refractive index, viscosity, and electrical conductivity of unsaturated solutions of sodium carbonate + poly(ethylene glycol) + water at four different temperature from 293.15 and 308.15 K. They correlated the experimental data on density, refractive index, and viscosity using Othmer's rule. Also the data on electrical conductivity were transformed to molar conductance and was correlated by group according to the percentage of PEG in each system.

A. Bhattacharjee [40] *et al.,* reported study of density and viscosity data for binary mixture of 1-alkyl-3-methylimidazolium alkylsulfates + water using ionic liquids such as 1-butyl-3-methylimidazolium hydrogensulfate, 1-butyl-3-methylimidazolium methylsulfate, 1-ethyl-3- methylimidazolium methylsulfate and 1-ethyl-3-methylimidazolium ethylsulfate The choice of the ionic liquids was compared with the effects of incrementing the alkyl chain length in the cation and the anion on the measured thermophysical properties behavior. They applied the Gardas and Coutinho group contribution methods to describe the pure component densities and viscosities. They also calculated and correlated excess molar volumes and viscosity deviations by Redlich–Kister polynomial expansions.

N. V. Sastry and I. R. Ravalji [41] measured density an speed of sound for six binary mixtures of water + room temperature ionic liquids (RTILs) of type $[C_n py][X]$ or $[C_n 4mpy][X]$ (where n = 6 or 8, [X] = Cl^- or Br^-) at temperature ranging from 308.15 and 318.15 K at atmospheric pressure. They calculated molar excess volumes excess speed of sound and excess isentropic compressibilities from the experimental data. They found that addition of a small amount of RTIL drastically increased the densities of water initially. The molar excess volume values for water + chloride-based RTILs were more negative across the composition compared to the mixtures with bromide-based RTILs; the negative magnitudes, however, decreased as the alkyl chain length and temperature increased. The negative molar excess volume values suggested the predominance of strong hetero interactions, increases with the increase in hydrocarbon chain length. Analysis of partial excess molar volumes or isentropic compressibilities and their standard transfer functions showed that the Br^- anion has the weakest water- RTIL interactions than Cl- anions.

S. K. Mylona and M. J. Assael [42] presented new reference correlations density and viscosity of squalane at high pressure. They found that in the case of the density, the correlation, based on the Tait equation from 273 to 473 K at pressure to 200 MPa. At 0.1 MPa, while in the case of the viscosity, two correlations were present, one a function of density and temperature, based on the Assael-Dymond model, and the other a function of temperature and pressure, based on a modified Vogel-Fulcher-Tammann equation.

S. M. Naeem and co-workers [43] measured densities, ultrasonic velocities and viscosities of binary liquid mixtures of cyclohexanone with benzyl benzoate, including pure liquids, at 308.15, 313.15 and 318.15 K temperature. Using the experimental results, molar volume, isentropic compressibility, intermolecular free length, acoustic impedance, internal pressure, enthalpy, Gibbs free energy of activation deviation properties of these including partial molar volumes and excess partial molar volumes, partial molar volume of the component at infinite dilution and excess partial molar volume at infinite dilution were calculated. They observed negative values of as well as positive values of all above parameters for all mixtures showed presence of strong-dipole-dipole interactions.

D.I. Sagdeev *et al.* [44] measured the density and viscosity of binary 1-hexene + 1-octene mixtures at 298 to 470 K temperature and pressure 196 MPa. The measurements were made for concentrations of 0.200, 0.429, and 0.692 mole fraction of 1-octene. The measured velocities were used to calculate excess molar volume and viscosity differences. They found that the values of excess molar volume for 1-hexene + 1-octene mixtures are negative, while the viscosity differences are mostly positive at all temperatures and pressure over the whole concentration range The measured densities and viscosities were also used to develop a Tait – type equation state and viscosity correlation model for liquid 1-hexene + 1-octene mixtures. Theoretically based Arrhenius-Andrade and Grunberg and Nisan type equations were used to represent the temperature and concentration dependences of the measured viscosities for liquid 1-hexene + 1-octene binary mixtures at atmospheric pressure.

J. Jacquemin *et al.* [45] measured densities and viscosities as a function of temperature for ionic liquids at temperatures up to 393 K and 388 K. They performed a qualitative analysis of the evolution of density and viscosity with cation and anion chemical structures

K. Rajagopal and G. R. Richi [46] determined values of density, viscosity and ultrasonic speed of paracetamol in aqueous methanol solution at different temperature. Using density values, they calculated apparent molal volume, partial molal volume, molar expansivity, isobaric thermal expansion coefficient, second derivative of infinite dilution of partial molal volume, ultrasonic speed, isentropic compressibility, change in isothermal compressibility, relative change in isentropic compressibility, apparent molal compressibility and partial molal compressibility

By using the viscosity data, they also calculated B-coefficient, variation of B-coefficient with temperature, free energy of activation per mole solvent and solute.They used this calculated parameters to predict the solute-solute, solute-solvent interactions, structure making and breaking ability of the drug and hydration property of the drug in aqueous methanol solution.

R. V. Entrolizo [47] measured densities and viscosities of aqueous ternary solutions of 2-amino-2-methyl-1-propanol with ethylenediamine (EDA) over a wide range of temperatures (303.15 to 343.15 K) and compositions. The obtained experimental data for density and viscosity were represented as a function of temperature and amine concentration using a Redlich-Kister-type and Vogel–Tammann–Fulcher equation, respectively. They found that the predicted data agrees well with the experimental data suggesting that the applied models was having satisfactory interpretation.

K. Saravanakumar and T. R. Kubendran [48] measured densities and viscosities of binary mixtures of 1-4 dioxane and benzene or chlorobenzene as function of mole fraction at different temperature at atmospheric pressure. They used these values to calculate excess molar volume, viscosity deviations, Grunberg-Nissan interaction constant. The viscosity values were fitted in McAllister, Krishnan-Laddha, Jouyban-Acree model. They found that the excess volume was correlated by using Redlich-Kister polynomial equation to obtain their coefficients and standard deviations. They also found that in every cases the experimental data correlated corresponding model very well. They also discussed the molecular interactions existing between the components. The molecular interactions existing between the components.

C. Ujjwala and co-workers [49] measured density and viscosity and ultrasonic velocity of 70 per cent DMF water and 70 per cent methanol water solutions of synthesized hybrid drug Ambroxol hydrochloride. They investigated molecular interactions at 300.15 K. They calculated various acoustical parameters like free path length, Raos molar constant, Wada constant, acoustic impendence, relative association and viscous relaxation time. They correlated these parameters with concentration and they observed good correlation. Increase or decrease in acoustical parameters with concentration showed existence of strong molecular interactions in solution.

A. A. Shaikh *et al.* [50] presented research paper for determination of the density and viscosity parameters were determined for ethanol +water mixed solvent systems. From these parameters they determined the relative density and relative viscosity for the solutions of Metformin Hydrochloride drug in ethanol + water mixed solvent systems. The data obtained was used to understand the effect of Metformin drug in ethanol-water mixed solvents. and solute solvent and solvent-solvent interaction in solution

G. A. Dopazo and co-workers [51] developed different artificial neural networks architectures to predict density, viscosity, and refractive index of binary and ternary mixture of ionic liquids. All neural network implemented were evaluated using RMS error. The individual model implemented showed greater values of R^2 and lower errors in RMSE. They found that result shown by different individuals artificial neural networks implemented are useful tools to predict density, viscosity and refractive index of binary and ternary mixture of ionic liquids with reasonable accuracy.

R. E. Pavai and S. Renuka [52] reported density, viscosity and ultrasonic velocity of L-threonine in aqueous potassium nitrate (0.04, 0.06 and 0.08M) solutions as the function of concentration of amino acid and electrolyte at 303 K. They used experimental data to estimate the adiabatic compressibility, change in adiabatic compressibility, relative change in adiabatic compressibility, apparent molal compressibility, apparent molal volume, limiting apparent molal volume and viscosity B-coefficient. They used these parameters to interpretate ion- ion and ion-solvent interactions

K. Anilkumar and Dr C. Srinivasu [53] measured sound velocity, densities of binary mixtures of 1,4-dioxane with 1-hexanol over the entire range of composition using Anton-Paar at temperatures from T=298.15, 303.15, 308.15, 313.15 and 318.15 K. Parameters like isentropic compressibility, excess molar volumes, excess sound velocity deduced. They found from experimental values that intermolecular interactions were present in the mixture. All the parameters were fitted to Redlich-Kister equation and their coefficient were obtained.

N. Santhi and J. Madhumitha [54] measured the ultrasonic velocity, density and viscosity for the binary liquid mixtures containing 2-(p-methoxy phenylsulphonyl)-4-methyl sydnone in Ethanol at 303.15K. From the data acoustic parameters such as adiabatic compressibility, free length, acoustic impedance, relative association, relaxation strength, Rao's constant (R), molar compressibility, relaxation time, Vander waal's constant, free volume and internal pressure was computed using the standard relations. They discussed the results in terms of molecular interactions.

U. K. Asema *et al.* [55] studied the effect of KCl concentration ranging from 2 per cent to 10 per cent on density and viscosity ethanol-water mixed solvent systems ranging from 5 per cent to 40 per cent. The data obtained was used to calculate the excess viscosity and B coefficient. They found that as the concentration of KCl increases the density and viscosity also increases it is due to solutet-solvent interaction. The KCl solute breaks the solvent-solvent molecules *i.e.* ethanol-water molecule due to which there is increase in density and viscosity of mixed solvent systems. The increase in excess viscosity which also gives idea about the dispersion types of forces exerts predominant behavior in these mixed solvent systems due to addition of KCl electrolyte.

E. Kartzmark [56] determined conductances, densities and viscosities of solutions of sodium nitrate in water in 20, 40, 60, 70, and 80 wt. per cent dioxane-water have measured at 25°C. They calculated apparent molar volumes. They found that there was no appreciable ion association is apparent in solutions whose dioxane content is less than 60 per cent by weight. They found a sharp increase beyond 60 per cent.

P. W. Derks and co-workers [57] assumed that the density, viscosity, and gas solubility data of Cullinane as a function of concentration when the values were reported as molalities. Using the corrected compositions, they concluded that the densities and viscosities determined agree well with those reported by Cullinane.

P. J. W. Derks *et al.* [58] determined densities and viscosities of aqueous solution containing piperazine (PZ) and N-methyldiethanolamine (MDEA) at (293.15 to 323.15) K. They kept concentration of MDEA constant at (1, 2, 3, and 4) mol $\cdot dm^{-3}$ by varying concentration of PZ (0 to 1) mol dm^{-3}.

G. S. Reddy and M. M. Reddy [59] determined the experimental densities and viscosities of Toluene with benzene at temperatures of 303.15, 308.15 and 313.15K using a vibrating-tube density meter and Oswald Viscometer. They calculated excess molar volumes and viscosity deviations. They discussed results in terms of the molecular interactions.

V. Gupta and co-workers [60] reported density, viscosity and ultrasonic velocity of aqueous acetonitrile solutions were measured as functions of concentration at 308.15K. The measured data was used to calculate various acoustical and thermodynamic parameters *viz.* adiabatic compressibility, intermolecular free length, acoustic impedance, relaxation time, Rao's constant, Wada's constant, free volume, absorption coefficient, Gibb's free energy, relative association and available volume. These parameters helped in elucidating the molecular association in the mixture. The variation of these parameters with concentration of solute indicated the nature interaction present in the binary mixture.

S. S. Kulkarni and U. V. Khadke [61] measured ultrasonic velocity, density and viscosity for the mixture of polyvinylidene fluoride (PVDF) in acetone and dimethylformamide (DMF) of various stoichiometric ratios at 300K. The acoustic parameters like adiabatic compressibility, intermolecular free path length, acoustic impedance, relative association, ultrasonic attenuation, and relaxation time were estimated using experimental data with well-known techniques. The variation of these acoustic parameters was explained in terms of solute-solvent molecular interaction in a polymer solution.

E. D. Dikio *et al.* [62] determined densities, viscosities of binary liquid mixtures of pyridine and some primary alcohols at 293.15, 303.15, 313.15 and 232.15 K temperature. From the experimental data, deviation in viscosity, excess molar volume, excess Gibbs free energy of activation, were determined. The deviations were correlated with Redlich-Kister polynomial equation. Other parameters like Grunberg-Nissan interaction and Modified Kendall-Monroe equation were used for quantitative analysis of interactions in the system.

M. T. Andreuccetti and co-workers [63] characterised black liquor samples from the evaporation unit process streams, and measured physical properties (density, viscosity and total solids content with concentration of sodium sulfate. They found that experimental data showed good reproducibility, and correlation between the physical properties measured and sodium sulphate content.

G. Bolat and co-workers [64] reported the dynamic viscosities of acetic acid + water, benzene, toluene, n-hexane and n-heptane at five temperatures in the range of 230 C to 460 C throughout the whole concentration range. The viscosity deviations were calculated from experimental data and fitted to Redlich-Kister polynomial function They analysed results in terms of molecular interactions.

G. S. Sanap *et al.* [65] determined density, viscosity and refractive index of drug molecule Metformin HCl (MFH) in aqueous solution at different concentrations have been determined. They analysed density, viscosity and refractive index data for the evaluation of apparent molar volume, molar excess volume, excess viscosity, specific refraction and molar refraction. They discovered that drug contains structure-making compound due to presence of hydrophobic hydration of drug molecules. They also found B-coefficients values positive showing solute-solvent interactions.

Z. P. Visak and co-workers [66] calculated density, viscosity, excess molar volumes, viscosity deviations for the ternary systems water + butyl acetate + methanol and water + ethyl propionate + methanol at 303.15 temperature and atmospheric pressure. They used Redlich-Kister type equation to correlate binary and ternary data. This equation was used to calculate the above referred properties along with bimodal curve.

G. S. Reddy and M.M. Reddy [67] determined the experimental densities and viscosities of Methyl Ethyl Ketone with Ethyl Benzene at different temperatures. From these data, they calculated excess molar volume and viscosity deviations. The computed quantities were fitted to the Redlich-Kister equation to derive the coefficient and estimate the standard error values. The result was discussed in terms of intermolecular interactions.

M. V. Rathnam *et al.* [68] measured densities and viscosities of binary mixtures of hexyl actate with chlorobenzene, bromobenzene, fluorobenzene and nitrobenzene at 298.15, 303.15, 308.15 and 313.15 K over entire composition range. Along with speed of sound at 303.15 K. From density and viscosity data they calculated values of excess molar volume, free energy of activation, while from speed of sound data they evaluated isentropic compressibility, intermolecular free length, internal pressure, free volume and specific acoustic impedance. They found that the excess molar volumes are negative for all composition range at all temperatures. They also correlate viscosity using Frenkel and McAllister models.

R. B. Dhake [69] measured densities, viscosities of some substituted heterocyclic drug (Ebastine) in ethanol in the temperature range 303K to 318K. The viscosity was determined by John–Dole equation. The data was used to calculate Gibb's free energy change, entropy change, and enthalphy change. He found that ΔG

and ΔH are negative and ΔS is positive which indicate the spontaneity of reaction according to thermodynamics.

J. P. Singh and R. Sharma [70] studied the density, viscosity and ultrasonic velocity of cholesteryl carbonate. From the experimental data various thermo acoustical parameters such as free volume, internal pressure, Rao's constant and Wada's constant were evaluated. Molecular interactions in aqueous cholesteryl oleyl carbonate in terms of thermo-acoustical parameters were also discussed.

K. Das and M. N. Roy [71] studied densities, viscosities, apparent molar volume, refractive index, of oxalate salts ($Li_2C_2O_4$, $Na_2C_2O_4$, $K_2C_2O_4$, $(NH_4)_2C_2O_4$) have been studied in different mass fraction of ascorbic acid in water (H_2O) at 298.15 K, respectively. Masson equation was used to find the extent of interaction in terms of limiting apparent molar volume by extrapolating to zero concentration and experimental slopes which interpreted the solute-solvent and solute-solute interactions respectively in the solutions Using the Jones–Dole equation, the viscosity data were analyzed to determine the viscosity A and B-coefficient, which interpreted the solute-solute and solute-solvent interaction respectively in the solutions. Molar refractions were calculated with the help of the Lorentz–Lorenz equation. The role of the solvent and the contribution of solute-solute and solute-solvent interactions to the solution complex was analysed through derived properties. The Gibbs energies of mixing for $K_2C_2O_4$-vitamin-C binary solids and liquids and solid-saturated $K_2C_2O_4$-vitamin-C-H_2O ternary liquids were modelled using asymmetric Margules treatments.

M. Asif and co-workers [72] determined the viscosity and density parameters by taking different concentrations of calcium chloride in different sets of Ethanol-Water mixed solvent systems at 298.5 K. The data was used to calculate the excess viscosity and B-coefficient values. From the viscosity and excess viscosity values the effect of electrolytic concentration was discussed and related to solute-solvent and solvent-solvent interactions.

R. R. Naik and co-workers [73] measured ultrasonic velocity, density, and viscosity in the binary mixture of paracetamol with water in concentration ranging from (0.1 to 0.0125 per cent) at 303 K and 313 K. They used measured value of density, ultrasonic velocity, and viscosity to estimate the acoustical parameters like adiabatic compressibility, relation time, acoustic impedance, free length, free volume, internal pressure and Wada's constant. The obtained results supported the complex formation, molecular association by intermolecular hydrogen bonding in the binary liquid mixtures.

P. G. Tello and co-workers [74] measured the densities and viscosities of concentrated aqueous solutions of polyethylene glycol (10-50 mass per cent) 277 to 298 K. It showed a linear variation with the polyethylene glycol concentration and differ from those of pure water at the same temperature. They also studied the viscosity of mixtures of polyethylene glycol + magnesium sulfate + water. A linear relationship exists between the viscosities of the aqueous solutions of polyethylene glycol and the concentration of magnesium sulphate was observed by them.

G. P. Borse *et al.* [75] reported the ultrasonic velocity, viscosity and density at 298.15 K in the binary mixture of methanol and ethanol with p-anisaldehyde over entire range of mole fraction. From the experimental data various acoustical parameters such as acoustic impedance, relative association, internal pressure, intermolecular free length, adiabatic compressibility and apparent molar volume were evaluated. The results were interpreted in terms of molecular interactions between components of the mixtures.

K. C. Patil and C. M. Dudhe [76] measured density, viscosity and ultrasonic velocity of aqueous solution of aminoglycoside antibiotic- Neomycin sulphate at temperatures 298.15, 303.15 and 308.15 K at various concentrations. From the experimental data, they calculated important acoustic parameters such as acoustic impedance, adiabatic compressibility, free length, relaxation time, internal pressure, absorption coefficient, free volume, Rao's constant, Wada's constant, cohesive energy, Gibb's free energy, relative association and Van der Waal's constant. These parameters was used to study the molecular interactions in aqueous neomycin solutions. They found that the classical absorption and relaxation time decreases with the rise in temperature, due to weakening of inetrmolecular hydrogen bondings.

V. D. Bhandakkar [77] measured the densities, viscosities and ultrasonic velocity of binary liquid mixture of acrolein with the solvent methanol. The observed experimental data was utilized to evaluate some of the derived thermodynamic parameters such as adiabatic compressibility, free length, free volume, internal pressure, relaxation time and acoustic impedance. The behaviour of these parameters with composition of the mixture was discussed in terms of molecular interactions between the components of liquids. The excess values of adiabatic compressibility free volume and free length were evaluated and discussed for F molecular interactions in the mixture.

E. Obermeier and co-workers [78] measured density at -10°C to 140°C while viscosity at -10°C to 81°C. Prandtl numbers of aqueous solution of di and triethylene glycol was determined.

K. Kaur and K. C. Juglan [79] calculated density, viscosity and ultrasonic velocity in binary mixtures of ethyl acetate and hexane at temperature 292 K. Using experimental values various acoustic parameters such as adiabatic compressibility, intermolecular free length, acoustic impedance, ultrasonic attenuation, available volume, free volume, internal pressure, Gibbs free energy and enthalpy were calculated. Theoretical values of ultrasonic velocity were computed using various models such as impedance relation, Nomoto's relation, Junjie's equation and Van-Deal and Vangeel. Further the chi-square test was also applied to check the relation between theoretical and experimental data. The non-linear variation in acoustical parameters showed that there is a complex formation and the deviation in experimental values from theoretical models proved the strong molecular interaction in the binary mixture.

C. S. Selvi *et al.* [80] have reported ultrasonic velocity, viscosity and density of 2-chlorobenzaldehyde with iodine in hexane ternary solution at 303K in different concentrations. From ultrasonic velocity, density and coefficient of viscosity, the acoustical parameters like adiabatic compressibility, free length, interaction parameter, Free volume were calculated. The results was used to discuss the nature, strength of intermolecular interactions, solute-solute and solute-solvent inetractions between the component.

B. Saini and co-workers [81] has reported the apparent molal volume and adiabatic compressibility of L-arginine hydrochloride in water and aqueous tetrahydrofuran at 303.15 K from density and sound measurements. The experimental data was used to evaluate various thermodynamic and acoustic parameters like partial molal volume at infinite dilution, partial molal adiabatic compressibility at infinite dilution, relative association, *etc.* The data was also investigated in terms of relative viscosity and B- coefficient of Jones – Dole equation.

4.3 Present Work

All of the above reported literature surveys pointed that density, viscosity and molar volume study of dihydropyrimidines and related compounds such as dihydropyrimidine carbonitriles, dihydropyrimidin-2(1H)-ones and dihydropyrimidine carboxylates has not been studied before. Hence in our present work we synthesized substituted dihydropyrimidine carbonitriles (4a-4e), dihydropyrimidin-2(1H)-ones (5a-5e) and dihydropyrimidine carboxylates (6a-6e) (Table:3.1, Chapter-3) by new developed protocol one pot efficient, solvent free, green method using ammonium chloride, different substituted aromatic aldehydes, respective 2,4 dicarbonyl compounds and urea/thiourea at 100°C. Synthesized compounds were purified by recrystallization method using proper and suitable solvent. The structure of compounds was confirmed by infrared spectroscopy (IR) and ^{1}H-NMR spectroscopic techniques. 60 per cent (v/v) dimethyl sulphoxide aqueous solution was prepared by using purified and distilled dimethyl sulphoxide and triple distilled water. Five molar concentrations of 0.002, 0.004, 0.006, 0.008 and 0.1 M were prepared for each compounds of dihydropyrimidine carbonitriles (4a-4e), dihydropyrimidin-2(1H)-ones (5a-5e) and dihydropyrimidine carboxylates (6a-6e) before the measurement of density and viscosity. After preparation of solutions, density, viscosity and relative viscosities of all the above solutions were determined at 298.15, 303.15, 308.15 and 313.15 K temperature by measuring the volume using pycnometer and efflux time by Ubbelohde viscometer as discussed in Chapter 4. From these measurement, other parameters such as relative viscosities, apparent molar volumes, limiting apparent molar volumes, Intercept value or Falkenhagen coefficient "A" and experimental slope or Jones-Dole coefficient "B" which predicts different types of interactions such as solute-solvent, solute-solute and solvent-solvent interactions present in the solution. Intercept value "A" and

Jones-Dole coefficient "B" are obtained from Masson [19] and Jones-Dole equation [24] as discussed earlier.

4.4 Result and Discussion

4.4.1 Density and Viscosity of Dihydropyrimidine Carbonitrile (4a-4e)

Density, viscosity and relative viscosities of dihydropyrimidine carbonitriles (4a-4e) of different molar concentration (0.002-0.01 mol dm^{-3}) in 60 per cent (v/v) dimethyl sulphoxide aqueous solution at 298.15, 303.15, 308.15, 313.15 K temperature as shown in Table 4.1. Apparent molar volumes $Ø_v$ was calculated by using densitometric values from Masson equation [19] as shown below

$$Ø_v = \frac{M}{\rho^o} - \frac{1000\,(\rho - \rho^o)}{C\rho^o}$$

where,

M = Molecular mass in grams

ρ = Density of solution in g cm^{-3}.

ρ^0 = Densities of solvent in g cm^{-3}

C = Molar concentration of solution.

The apparent molar volume data are shown in Table 4.1 for dihydropyrimidine carbonitriles (4a-4e) under study. From the above values it is found that there is slight increase in density with increase in concentration of solution of dihydropyrimidine carbonitriles at 298.15, 303.15, 308.15 and 313.15 K temperature. It was also observed that there was decrease in density with increase in temperature along with increase in apparent molar volume with increase in concentration and temperature for dihydropyrimidine carbonitriles. The following equation gives relation between apparent volume and concentration

$$Ø_v = Ø_v^0 + S_v\,C^{1/2}$$

where,

$Ø_v$ = Limiting apparent molar volume or apparent molar volume at infinite dilution,

$Ø_v^0$ = Limiting value of apparent molar volume (*i.e.* apparent molar volume at infinite dilution) which gives an idea about solute-solvent interactions

S_v = Experimental slope or limiting slope and determines extent of solute-solute interactions.

Values of limiting apparent molar volume "$Øv^0$" of dihydropyrimidine carbonitriles studied at 298.15, 303.15, 308.15 and 313.15 K temperature was determined by plotting graph of $Ø_v$ against √C as shown in Figure 4.1 to infinite dilution along with the value of "S_v" as the slope with correlation coefficient 0.78470-0.9059. "$Ø_v^0$" and "S_v" values for dihydropyrimidine carbonitriles are shown in Table 4.2. highly positive "S_v" values interprets strong solute-solute interactions and decreases with increase in temperature for all dihydropyrimidine carbonitriles compounds indicating that with increase in temperature solute-solute interactions become weak *i.e.* are stronger at low temperature (298.15 K) and weaker at high temperature (313.15 K). This impacts that with increase in temperature, the interactions diminishes due to increase in internal energy of solute molecules present in the solution.

Table 4.1: Density (ρ), Apparent Molar Volume ($Ø_v$), Viscosity (η) and Relative Viscosity ($η_r$) of Dihydropyrimidine Carbonitriles (4a-4e) in 60 per cent Aqueous DMSO at different Temperatures

Comp	*Conc (mol dm⁻³)*	*√C*	*Density "ρ" (g cm⁻³)*	*$Ø_v$ (cm³.mol⁻¹)*	*Viscosity "η" (cP)*	*Relative Viscosity "$η_r$"*
298.15 K						
4a	0.002	0.0447	1.08527	-1086.4143	3.70105	1.098043
	0.004	0.0632	1.08547	-483.6206	3.70728	1.099891
	0.006	0.0775	1.08562	-274.9909	3.71334	1.101689
	0.008	0.0894	1.08585	-179.9143	3.71968	1.103569
	0.01	0.1	1.08623	-136.7256	3.72653	1.105602
4b	0.002	0.0447	1.08925	-2907.7287	3.73009	1.106657
	0.004	0.0632	1.08868	-1207.8968	3.72981	1.106576
	0.006	0.0775	1.08978	-898.4166	3.74318	1.110542
	0.008	0.0894	1.08982	-621.2701	3.74777	1.111902
	0.01	0.1	1.08994	-462.3727	3.75297	1.113446
4c	0.002	0.0447	1.08736	-2112.3182	3.70793	1.100084
	0.004	0.0632	1.08772	-971.3985	3.71368	1.101791
	0.006	0.0775	1.08815	-632.6639	3.72071	1.103874
	0.008	0.0894	1.08893	-503.7138	3.7255	1.105296
	0.01	0.1	1.08912	-371.8382	3.73232	1.10732
4d	0.002	0.0447	1.09076	-3580.7328	3.79084	1.124682
	0.004	0.0632	1.09188	-1922.4729	3.79696	1.126498
	0.006	0.0775	1.09385	-1500.5943	3.79766	1.126705
	0.008	0.0894	1.09456	-1144.1531	3.80988	1.130329
	0.01	0.1	1.09586	-984.7939	3.81716	1.13249
4e	0.002	0.0447	1.08957	-3044.9162	3.76793	1.117885
	0.004	0.0632	1.09102	-1737.7086	3.77488	1.119947
	0.006	0.0775	1.09156	-1161.8597	3.78107	1.121783
	0.008	0.0894	1.09189	-849.6850	3.79258	1.125198
	0.01	0.1	1.09387	-814.8107	3.7985	1.126955

Comp	Conc (mol dm^{-3})	√C	Density "ρ" (g cm^{-3})	$Ø_v$ (cm^3.mol^{-1})	Viscosity "η" (cP)	Relative Viscosity "$η_r$"
303.15 K						
4a	0.002	0.0447	1.08395	-990.4287	3.05902	1.0227
	0.004	0.0632	1.08418	-442.50247	3.06485	1.02465
	0.006	0.0775	1.08434	-249.0714	3.07062	1.02658
	0.008	0.0894	1.08456	-159.29164	3.07766	1.028938
	0.1	0.1	1.08472	-99.8752	3.08367	1.03094
4b	0.002	0.0447	1.08769	-2702.6402	3.11534	1.04153
	0.004	0.0632	1.08934	-1618.3474	3.12118	1.04348
	0.006	0.0775	1.08978	-1070.4212	3.12845	1.04591
	0.008	0.0894	1.09089	-873.90762	3.13354	1.04761
	0.01	0.1	1.09112	-674.6197	3.13925	1.04952
4c	0.002	0.0447	1.08684	-2019.9361	3.08699	1.03205
	0.004	0.0632	1.08912	-971.3985	3.09328	1.03415
	0.006	0.0775	1.08942	-632.6639	3.09945	1.03622
	0.008	0.0894	1.09068	-503.7138	3.105	1.03807
	0.01	0.1	1.09072	-371.8382	3.11095	1.04006
4d	0.002	0.0447	1.08885	-3214.5004	3.17766	1.06236
	0.004	0.0632	1.0897	-1677.0703	3.18365	1.06437
	0.006	0.0775	1.09194	-1378.832	3.18974	1.06640
	0.008	0.0894	1.09211	-990.4286	3.19609	1.06853
	0.01	0.1	1.09312	-835.0673	3.20168	1.070340
4e	0.002	0.0447	1.08836	-3001.8033	3.14994	1.05310
	0.004	0.0632	1.08948	-1640.0795	3.15784	1.05574
	0.006	0.0775	1.08998	-1090.612	3.16396	1.05770
	0.008	0.0894	1.09134	-915.2911	3.17019	1.05987
	0.01	0.1	1.09215	-759.23614	3.1768	1.06208
308.15 K						
4a	0.002	0.0447	1.07465	-564.7822	2.62063	1.01433
	0.004	0.0632	1.07478	-205.9684	2.62645	1.01659
	0.006	0.0775	1.0752	-131.4097	2.63215	1.01879
	0.008	0.0894	1.07565	-97.6253	2.63959	1.02167
	0.1	0.1	1.07585	-54.0551	2.64559	1.02340
4b	0.002	0.0447	1.07799	-2103.9535	2.70604	1.04739
	0.004	0.0632	1.07837	-1025.1822	2.71251	1.04990
	0.006	0.0775	1.07871	-659.3786	2.71889	1.05236
	0.008	0.0894	1.07875	-441.5273	2.72451	1.05454
	0.01	0.1	1.07878	-309.8846	2.7301	1.05671

Comp	Conc (mol dm^{-3})	√C	Density "ρ" (g cm^{-3})	$Ø_v$ (cm^3.mol^{-1})	Viscosity "η" (cP)	Relative Viscosity "$η_r$"
4c	0.002	0.0447	1.07659	-1436.6531	2.65843	1.02897
	0.004	0.0632	1.078	-924.0620	2.66743	1.03245
	0.006	0.0775	1.07943	-756.3049	2.6765	1.03596
	0.008	0.0894	1.07796	-334.5822	2.67837	1.03668
	0.01	0.1	1.07654	-86.2085	2.68035	1.03745
4d	0.002	0.0447	1.07876	-2438.0697	2.73005	1.05669
	0.004	0.0632	1.07936	-1231.1506	2.7371	1.05941
	0.006	0.0775	1.07985	-811.7580	2.74387	1.06203
	0.008	0.0894	1.08035	-603.2265	2.75067	1.06466
	0.01	0.1	1.08096	-488.3595	2.75775	1.06741
4e	0.002	0.0447	1.07845	-2307.5920	2.71823	1.05211
	0.004	0.0632	1.07885	-1126.30250	2.72476	1.05464
	0.006	0.0775	1.07963	-791.56493	2.73226	1.05754
	0.008	0.0894	1.07979	-551.9674	2.73819	1.05983
	0.01	0.1	1.07982	-396.0931	2.74379	1.06200
313.15 K						
4a	0.002	0.0447	1.06656	-296.5827	2.27311	1.01321
	0.004	0.0632	1.06723	-198.0347	2.28	1.01628
	0.006	0.0775	1.06745	-94.7939	2.28594	1.01893
	0.008	0.0894	1.06769	-45.5198	2.29193	1.02160
	0.1	0.1	1.06818	-39.4192	2.29846	1.02451
4b	0.002	0.0447	1.06751	-725.0322	2.32984	1.03850
	0.004	0.0632	1.06763	-274.5267	2.33557	1.04105
	0.006	0.0775	1.06785	-140.0008	2.34153	1.04371
	0.008	0.0894	1.06806	-71.5647	2.34746	1.04635
	0.01	0.1	1.06842	-44.5813	2.35373	1.04915
4c	0.002	0.0447	1.06685	-400.2928	2.30106	1.02567
	0.004	0.0632	1.06715	-146.8835	2.30718	1.02840
	0.006	0.0775	1.06779	-115.5984	2.31404	1.03145
	0.008	0.0894	1.06812	-63.5870	2.32022	1.03421
	0.01	0.1	1.06855	-41.7656	2.32663	1.03707
4d	0.002	0.0447	1.06785	-859.7145	2.35247	1.04859
	0.004	0.0632	1.06812	-364.6278	2.35854	1.05129
	0.006	0.0775	1.06831	-187.0850	2.36443	1.05392
	0.008	0.0894	1.06862	-112.3917	2.37059	1.05666
	0.01	0.1	1.06891	-65.6987	2.37671	1.05939
4e	0.002	0.0447	1.06758	-747.0881	2.34093	1.04344

Comp	Conc (mol dm^{-3})	$\sqrt{C}$	Density "ρ" (g cm^{-3})	$Ø_v$ (cm^3.mol^{-1})	Viscosity "η" (cP)	Relative Viscosity "$η_r$"
	0.004	0.0632	1.06744	-219.1521	2.3461	1.04575
	0.006	0.0775	1.06763	-94.7938	2.35198	1.04837
	0.008	0.0894	1.06854	-43.1735	2.35807	1.05108
	0.01	0.1	1.06871	-61.0059	2.36532	1.05431

The viscosity data for all dihydropyrimidine carbonitriles under investigation were analyzed by Jones-Dole equation [24] as shown below

$$\frac{\eta_r - 1}{\sqrt{C}} = A + B\sqrt{C}$$

where

η_r = Relative viscosity

C = Molar concentration of solute.

The Falkenhagen constant or viscosity coefficient "A" which explains the nature and extent of solute-solute interactions are evaluated by plotting graph between $\eta_r - 1/\sqrt{C}$ against $\sqrt{C}$ at 298.15, 303.15, 308.15 and 313.15 K temperature as shown in Figure 4.2 at infinite dilution. Values of Jones-Dole or viscosity B-coefficient are determined from slope having correlation coefficient ranging between 0.7757-0.9724 at 298.15, 303.15, 308.15, 313.15 K temperature. The values of "A" and "B" for dihydropyrimidine carbonitriles are shown in Table 4.3. Positive "A" values reveals solute-solute interaction was found to be stronger at lower temperature (298.15 K) and becomes weaker as temperature increases *i.e.* higher temperature (313.15 K).

Hence from above apparent molar volume and Jones-Dole coefficients studies, we have concluded presence of strong solute-solute interactions in compounds of dihydropyrimidine carbonitriles and were observed to follow the trends as : 4d > 4c > 4b > 4c > 4a.

Table 4.2: Limiting Apparent Molar Volume "$Ø_v$" (cm^3 mol^{-1}) and Semi Empirical Parameters (Experimental slope "S_v") of Dihydropyrimidine Carbonitriles at different Temperature

Comp.	298.15 K		303.15 K		308.15 K		313.15 K	
	S_v	$Ø_v$	S_v	$Ø_v$	S_v	$Ø_v$	S_v	$Ø_v$
4a	16801	-1691.9	15692	-1564.7	8661.3	-860.14	4969	-507.41
4b	41876	-4359.2	36184	-4100.8	31611	-3278.0	11957	-1147.6
4c	30154	-3179.1	28679	-3050.1	23974	-2505.0	6112.1	-611.87
4d	45362	-5227.5	41307	-4716.1	34429	-3695.8	14013	-1368.5
4e	40601	-4565.8	39635	-4453.0	33333	-3533.8	12026	-1134.7

Figure 4.1: Graph of √C Vs $\emptyset_v$ for Dihydropyrimidine Carbonitriles (4a-4e) at different Temperatures

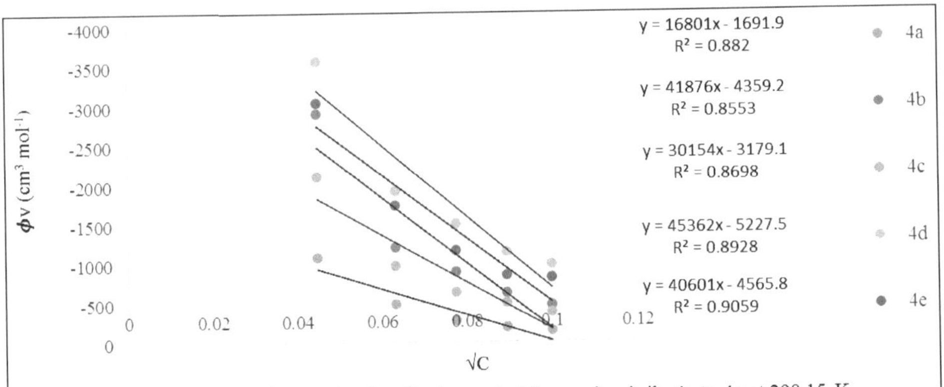

Figure:Plot of √C Vs ϕv for dihydropyrimidine carbonitrile 4a to 4e at 298.15 K

298.15 K

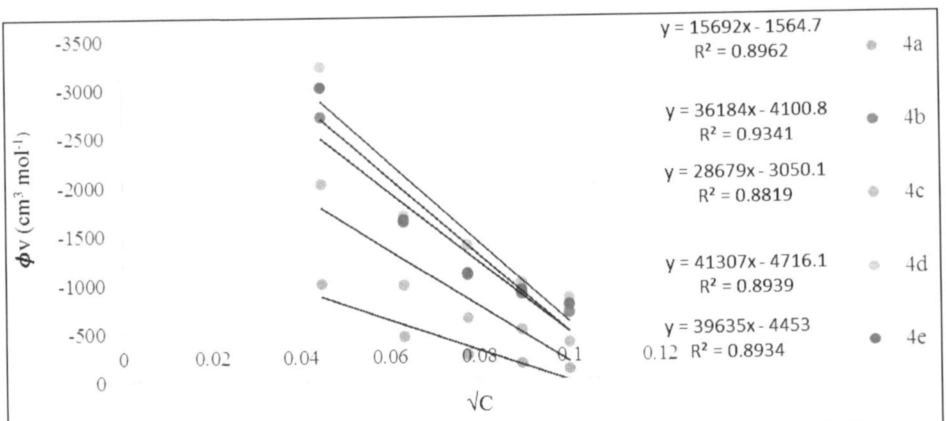

Figure:Plot of √C Vs ϕv for dihydropyrimidine carbonitrile 4a to 4e at 303.15 K

303.15 K

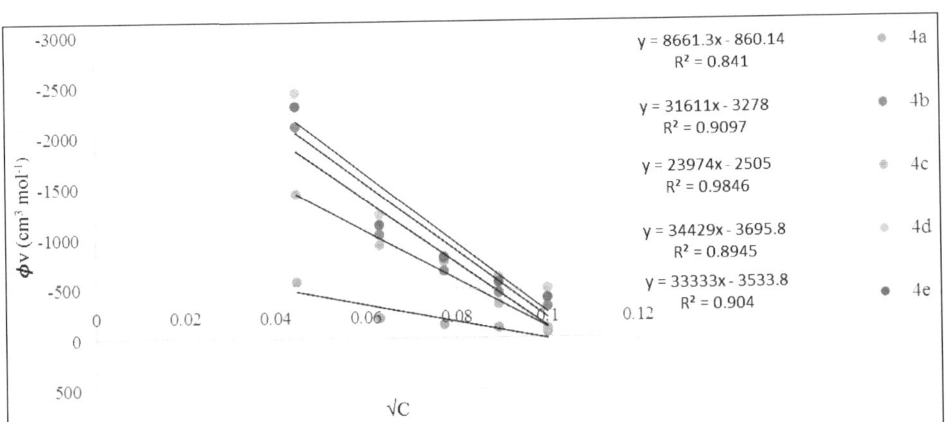

Figure:Plot of √C Vs ϕv for dihydropyrimidine carbonitrile 4a to 4e at 308.15 K

308.15 K

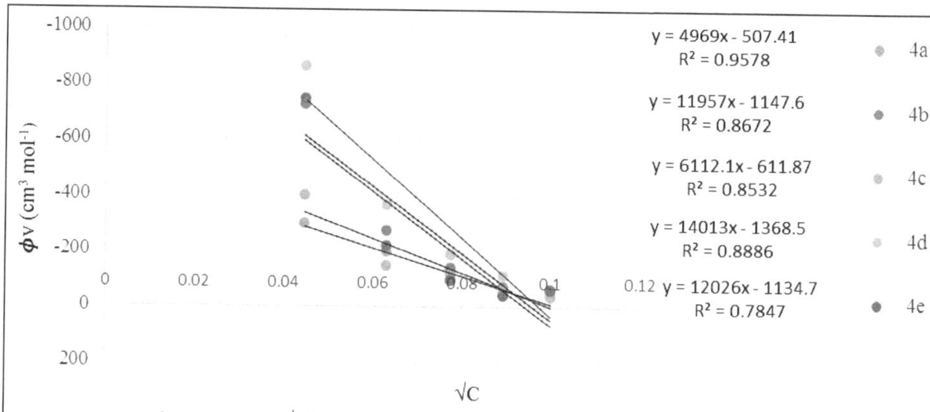

Figure:Plot of √C Vs ϕv for dihydropyrimidine carbonitrile 4a to 4e at 313.15 K

313.15 K

Table 4.3: Falkenhagen Coefficient ("A" $dm^{3/2}\ mol^{-1/2}$) and Jones-Dole Coefficient ("B" $dm^3\ mol^{-1}$) for Dihydropyrimidine Carbonitriles at different Temperatures

Comp.	*298.15 K*		*303.15 K*		*308.15 K*		*313.15 K*	
	A	*B*	*A*	*B*	*A*	*B*	*A*	*B*
4a	2.9793	-20.269	0.6378	-3.5101	0.3667	-1.4034	0.3248	-0.9078
4b	3.2336	-22.101	1.2246	-7.6998	1.3978	-8.77	1.1097	-6.5745
4c	3.0451	-20.772	0.9323	-5.6314	0.844	-4.8302	0.706	-3.595
4d	3.7998	-26.124	1.8696	-12.293	1.6741	-10.579	1.4218	-8.7711
4e	3.5773	-24.309	1.5799	-10.123	1.5379	-9.68	1.259	-7.6338

4.4.2 Density and Viscosity of Dihydropyrimidin-2(1H)-ones (5a-5e)

Density, viscosity and relative viscosities for dihydropyrimidin-2(1H)-ones were measured in mol dm^{-3} by using 60 per cent (v/v) aqueous dimethyl sulphoxide at 298.15, 303.15, 308.15 and 313.15 K temperature and presented in Table 4.4. From the density values, Øv were evaluated by using equation 5.2 and the data were tabularized in Table 4.4. We observed that the densities of solution slightly increasing with increase in molar concentration at 298.15, 303.15, 308.15 and 313.15 K and values of density decreases with increase in temperature at particular temperature. Also, with respect to temperature and concentration, densities, apparent molar volumes for all compounds of dihydropyrimidin-2(1H)-ones under observation increases.

Apparent molar volume has been determined by using equation 5.3 as discussed earlier. The $Øv^0$ values of dihydropyimidin-2(1H)-ones studied at 298.15, 303.15, 308.15 and 313.15 K temperature were estimated from plotting graph

Figure 4.2: Graph of $\eta_r - 1/\sqrt{C}$ Vs $\sqrt{C}$ for Dihydropyrimidine Carbonitriles (4a-4e) at different Temperatures

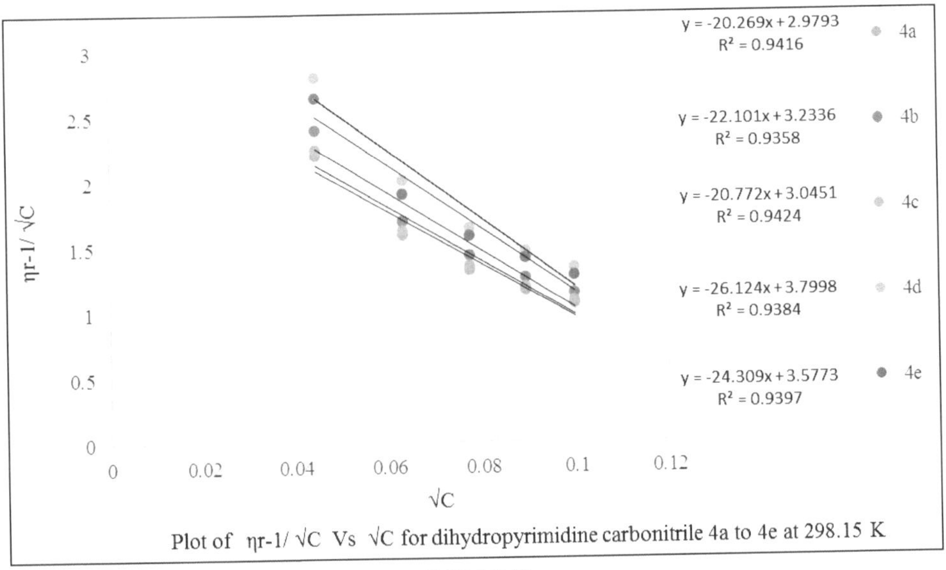

298.15 K

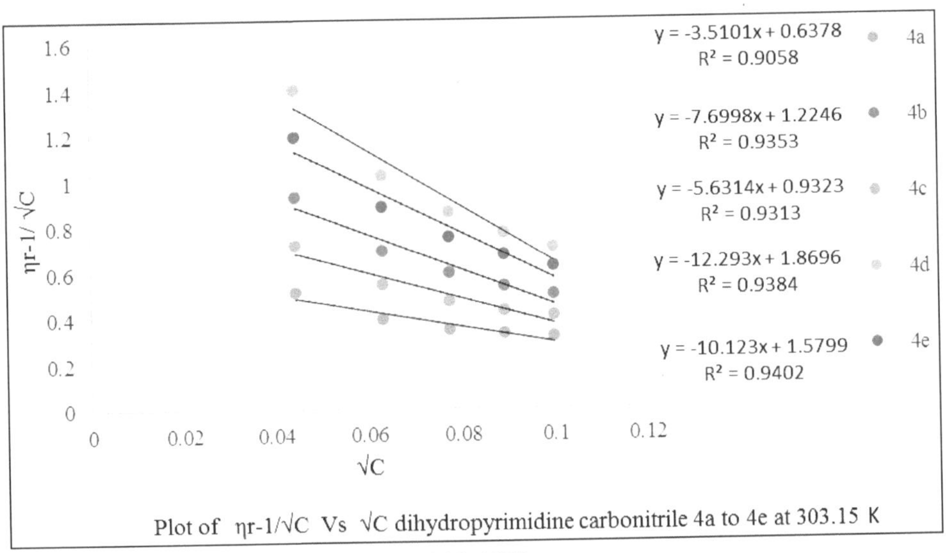

303.15 K

of "Øv" against $\sqrt{C}$ (Figure 4.3) at infinite dilution while slope of graph represent values of "S_v" (limiting slope) with correlation coefficient 0.8165-0.9775 as shown in Table 4.5. Large positive "S_v" values interpreted exixtance of strong solute-solute interactions and found to decrease with increase in temperature from 298.15 to 313.15 K temperature for all compounds of dihydropyrimidin-2(1H)-ones which

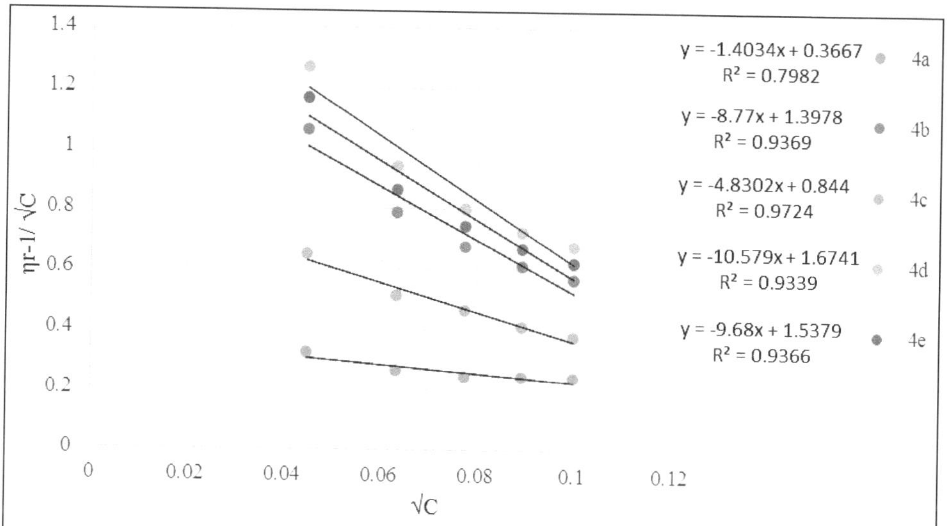

Plot of ηr-1/√C Vs √C dihydropyrimidine carbonitrile 4a to 4e at 308.15 K

308.15 K

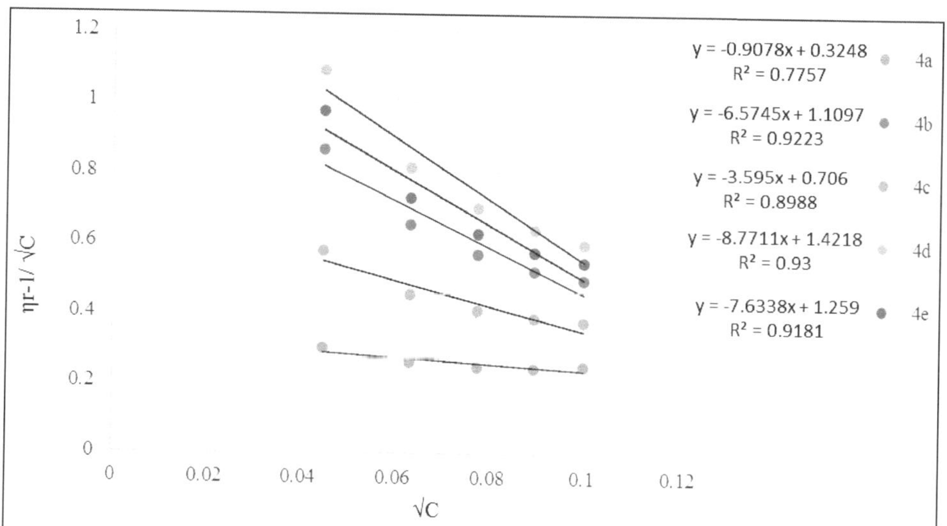

Plot of ηr-1/ √C Vs √C for dihydropyrimidine carbonitrile 4a to 4e at 313.15 K

313.15 K

suggested that the solute-solute interactions are strong at lower temperature (298.15 K) and weaker at elevated or high temperature (313.15 K).

The viscosity data for all compounds of dihydropyrimidin-2(1H)-ones under investigation were interpreted by Jones-Dole equation [24]. The viscosity coefficient or Falkenhagen coefficient "A" values of compounds of dihydropyrimidin-2(1H)-ones in the system were studied at 298.15, 303.15, 308.15, 313.15 K temperature by plotting graph of $\eta_r - 1/\sqrt{C}$ Vs $\sqrt{C}$ at different temperatures as shown in Figure 4.6. to infinite dilution along with Jones-Dole coefficient "B" values from the slope having correlation coefficient 0.7621-0.975. "A" and "B" values for dihydropyrimidin-2(1H)-ones are summarized in table 5.6. "A" with positive values clearly indicates presence of solute-solute interactions and found to decrease with temperature which clearly predicts solute-solute interactions which are stronger at lower temperature (298.15 K) while weaker at elevated or high temperature (313.15 K). Hence in our present study it is estimated that solute-solute interactions among the compounds from dihydropyrimidin-2(1H)-ones follow the order 5d > 5e > 5b >5c > 5a.

Table 4.4: Density (ρ), Apparent Molar Volume ($\varnothing_v$), Viscosity (η) and Relative Viscosity (η_r) of Dihydropyrimidin-2(1H)-ones (5a-5e) in 60 per cent Aqueous DMSO at different Temperatures

Comp	*Conc (mol dm⁻³)*	*√C*	*Density "ρ" (g cm⁻³)*	*$\varnothing_v$ (cm³.mol⁻¹)*	*Viscosity "η" (cP)*	*Relative Viscosity "η_r"*
			298.15 K			
5a	0.002	0.0447	1.08381	-396.4017	3.43019	1.01768
	0.004	0.0632	1.08465	-278.5734	3.43840	1.02012
	0.006	0.0775	1.08525	-202.3444	3.44585	1.02233
	0.008	0.0894	1.08565	-141.1343	3.45267	1.02435
	0.01	0.1	1.08625	-122.8847	3.46013	1.02656
5b	0.002	0.0447	1.08497	-915.1277	3.45050	1.02371
	0.004	0.0632	1.08653	-695.6789	3.46102	1.02683
	0.006	0.0775	1.0878	-577.878	3.47062	1.02968
	0.008	0.0894	1.08856	-460.0838	3.47861	1.03205
	0.01	0.1	1.089	-359.8450	3.48559	1.03412
5c	0.002	0.0447	1.08432	-600.1044	3.44289	1.02145
	0.004	0.0632	1.08556	-456.8710	3.45238	1.02427
	0.006	0.0775	1.08625	-324.4428	3.46013	1.02656
	0.008	0.0894	1.08715	-282.4791	3.46855	1.02906
	0.01	0.1	1.08818	-269.3105	3.47740	1.03169
5d	0.002	0.0447	1.0867	-1704.5334	3.53932	1.05006
	0.004	0.0632	1.0877	-956.1967	3.54814	1.05268
	0.006	0.0775	1.08876	-715.9893	3.55030	1.05332
	0.008	0.0894	1.08897	-497.7295	3.55543	1.05484
	0.01	0.1	1.08964	-409.2695	3.56108	1.05652
5e	0.002	0.0447	1.08591	-1338.7000	3.51455	1.04271
	0.004	0.0632	1.08662	-705.84095	3.52240	1.04504
	0.006	0.0775	1.0881	-613.4451	3.53006	1.04731
	0.008	0.0894	1.08833	-422.9000	3.53256	1.04806
	0.01	0.1	1.08853	-305.8015	3.53877	1.0499

Comp	*Conc (mol dm^{-3})*	*√C*	*Density "ρ" (g cm^{-3})*	*$Ø_v$ (cm^3.mol^{-1})*	*Viscosity "η" (cP)*	*Relative Viscosity "$η_r$"*
303.15 K						
5a	0.002	0.0447	1.08236	-238.2204	3.02160	1.01019
	0.004	0.0632	1.08312	-181.0696	3.02926	1.01275
	0.006	0.0775	1.08356	-112.6984	3.03603	1.01502
	0.008	0.0894	1.08415	-95.8522	3.04323	1.01742
	0.01	0.1	1.08456	-69.0987	3.04993	1.01966
5b	0.002	0.0447	1.08314	-581.7715	3.04040	1.01647
	0.004	0.0632	1.08383	-328.1077	3.04788	1.01897
	0.006	0.0775	1.08455	-248.1769	3.05545	1.02151
	0.008	0.0894	1.08459	-129.6062	3.06111	1.0234
	0.01	0.1	1.08562	-150.0159	3.06824	1.02578
5c	0.002	0.0447	1.08282	-419.0124	3.02842	1.01247
	0.004	0.0632	1.08352	-241.6419	3.03592	1.01498
	0.006	0.0775	1.08426	-188.6835	3.04354	1.01753
	0.008	0.0894	1.08512	-176.0758	3.05150	1.02019
	0.01	0.1	1.08574	-146.3168	3.05880	1.02263
5d	0.002	0.0447	1.08462	-1256.3895	3.08893	1.0327
	0.004	0.0632	1.08445	-461.7366	3.09399	1.03439
	0.006	0.0775	1.08495	-300.1180	3.09859	1.03593
	0.008	0.0894	1.08534	-206.5931	3.10540	1.03821
	0.01	0.1	1.08605	-180.0708	3.11172	1.04032
5e	0.002	0.0447	1.08385	-899.4292	3.05902	1.0227
	0.004	0.0632	1.08415	-391.4542	3.06541	1.02484
	0.006	0.0775	1.08486	285.3217	3.07212	1.02708
	0.008	0.0894	1.08512	-180.2373	3.07925	1.02947
	0.01	0.1	1.08586	-161.5754	3.08691	1.03202
308.15 K						
5a	0.002	0.0447	1.07386	-181.5190	2.60771	1.00933
	0.004	0.0632	1.07465	-160.1924	2.61513	1.0122
	0.006	0.0775	1.07515	-108.0376	2.62185	1.01481
	0.008	0.0894	1.07558	-73.8053	2.62841	1.01734
	0.01	0.1	1.07589	-42.0821	2.63467	1.01977
5b	0.002	0.0447	1.07436	-396.5591	2.62543	1.01619
	0.004	0.0632	1.07516	-261.7788	2.63289	1.01908
	0.006	0.0775	1.07569	-174.6745	2.63969	1.02171
	0.008	0.0894	1.07612	-119.4726	2.64626	1.02425
	0.01	0.1	1.07635	-67.7117	2.65233	1.0266

Comp	Conc (mol dm^{-3})	$\sqrt{C}$	Density "ρ" (g cm^{-3})	$Ø_v$ (cm^3.mol^{-1})	Viscosity "η" (cP)	Relative Viscosity "$η_r$"
5c	0.002	0.0447	1.07414	-279.84345	2.61389	1.01172
	0.004	0.0632	1.07468	-135.0288	2.62071	1.01436
	0.006	0.0775	1.07544	-120.9300	2.62763	1.01704
	0.008	0.0894	1.07596	-85.9211	2.63485	1.01983
	0.01	0.1	1.07635	-52.7999	2.64205	1.02262
5d	0.002	0.0447	1.07585	-1081.8167	2.64559	1.02399
	0.004	0.0632	1.07598	-443.0499	2.65142	1.02625
	0.006	0.0775	1.07612	-231.6809	2.65611	1.02807
	0.008	0.0894	1.07656	-160.9459	2.66239	1.03049
	0.01	0.1	1.07685	-104.5251	2.66856	1.03289
5e	0.002	0.0447	1.0751	-731.3903	2.63824	1.02115
	0.004	0.0632	1.07546	-320.9598	2.64382	1.02331
	0.006	0.0775	1.07602	-215.2159	2.65056	1.02592
	0.008	0.0894	1.07633	-133.2194	2.65730	1.02853
	0.01	0.1	1.07652	-72.8376	2.66377	1.03103
313.15 K						
5a	0.002	0.0447	1.06624	-131.8981	2.26149	1.00803
	0.004	0.0632	1.06695	-117.1003	2.26847	1.01114
	0.006	0.0775	1.06745	-79.3183	2.27500	1.01406
	0.008	0.0894	1.06789	-53.3881	2.28141	1.01691
	0.01	0.1	1.06815	-20.9360	2.28744	1.0196
5b	0.002	0.0447	1.06644	-208.3904	2.27285	1.0131
	0.004	0.0632	1.06728	-177.1679	2.28011	1.01633
	0.006	0.0775	1.06758	-82.2904	2.28708	1.01944
	0.008	0.0894	1.06825	-78.2598	2.29313	1.02214
	0.01	0.1	1.06852	-38.2992	2.29919	1.02484
5c	0.002	0.0447	1.06635	-151.1385	2.26719	1.01057
	0.004	0.0632	1.06714	-129.3015	2.27370	1.01348
	0.006	0.0775	1.06767	-81.3518	2.28047	1.01649
	0.008	0.0894	1.06796	-29.2203	2.28704	1.01942
	0.01	0.1	1.0686	-30.7908	2.29358	1.02234
5d	0.002	0.0447	1.06812	-986.9223	2.29285	1.02201
	0.004	0.0632	1.06828	-401.9520	2.29867	1.02461
	0.006	0.0775	1.0681	-153.7770	2.30376	1.02687
	0.008	0.0894	1.06866	-116.5060	2.30897	1.0292
	0.01	0.1	1.06876	-50.9698	2.31570	1.0322
5e	0.002	0.0447	1.06695	-436.9287	2.28487	1.01846

Comp	Conc (mol dm^{-3})	$\sqrt{C}$	Density "ρ" (g cm^{-3})	$Ø_v$ (cm^3.mol^{-1})	Viscosity "η" (cP)	Relative Viscosity "$η_r$"
	0.004	0.0632	1.06756	-232.0734	2.29077	1.02108
	0.006	0.0775	1.0677	-90.2681	2.29742	1.02405
	0.008	0.0894	1.06813	-53.3881	2.30382	1.0269
	0.01	0.1	1.06866	-40.6457	2.31023	1.02976

Figure 4.3: Graph of Øv Vs √C for Dihydropyrimidine-2(1H)-ones (5a-5e) at different Temperatures

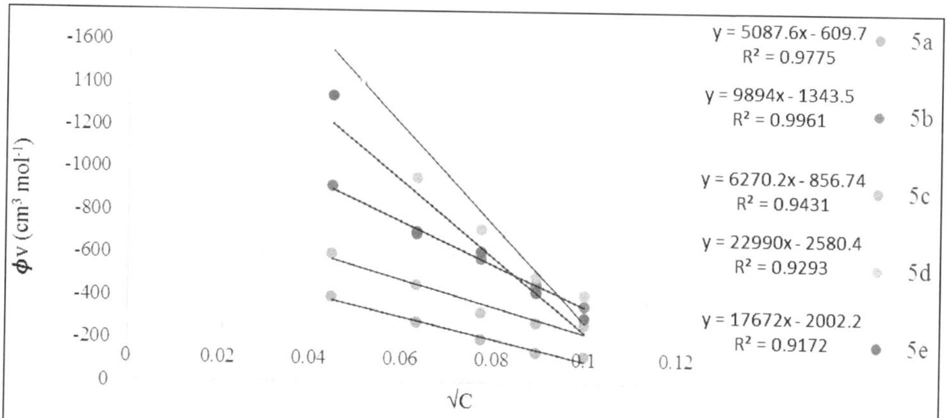

Figure: Plot of √C Vs ϕv for dihydropyrimidin-2(1H)-one 5a to 5e at 298.15 K

298.15 K

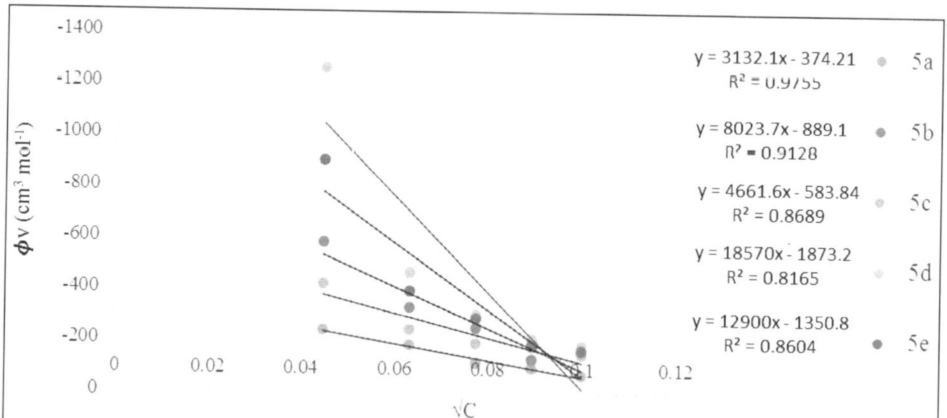

Figure:Plot of √C Vs ϕv for dihydropyrimidin-2(1H)-one 5a to 5e at 303.15 K

303.15 K

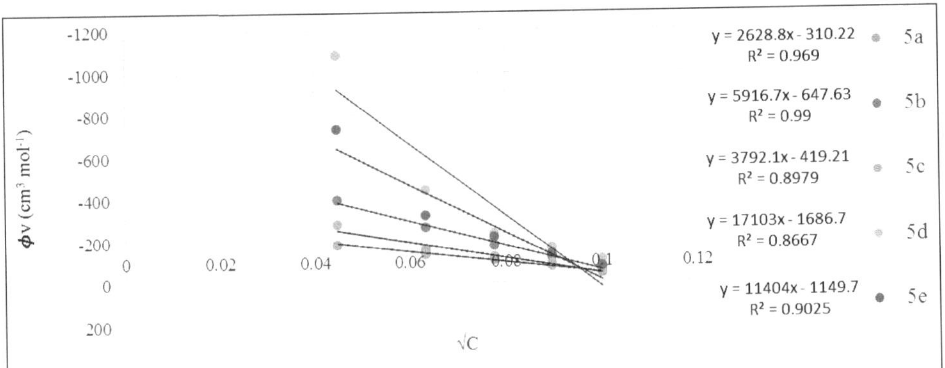

Figure: Plot of √C Vs ϕv for dihydropyrimidin-2(1H)-one 5a to 5e at 308.15 K

308.15 K

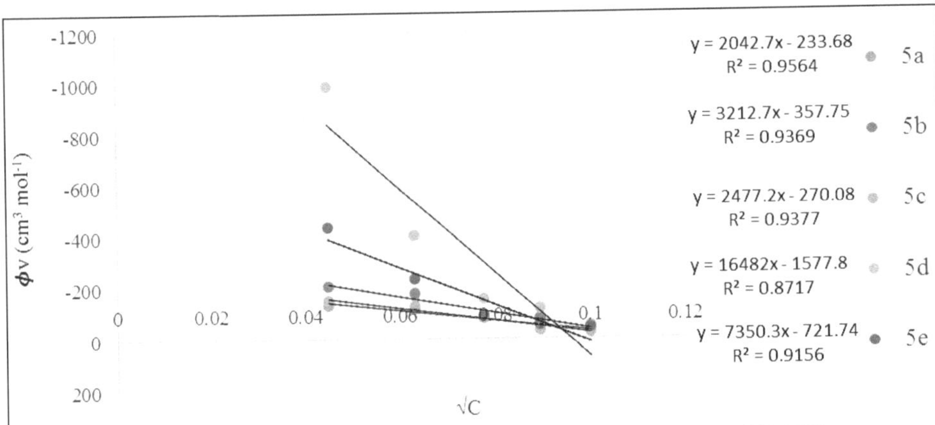

Figure:Plot of √C Vs ϕv for dihydropyrimidin-2(1H)-one 5a to 5e at 313.15 K

313.15 K

Table 4.5: Limiting Apparent Molar Volume "$Ø_v^0$" (cm^{-3} mol^{-1}) and Semi Empirical Parameter (Experimental slope "S_v") of Dihydropyrimidin-2(1H)-ones at different Temperatures

Comp.	*298.15 K*		*303.15 K*		*308.15 K*		*313.15 K*	
	S_v	$Ø_v^0$	S_v	$Ø_v^0$	S_v	$Ø_v^0$	S_v	$Ø_v^0$
5a	5087.6	-609.7	3132.1	-374.21	2628.8	-310.22	2042.7	-233.68
5b	9894	-1343.5	8023.7	-889.1	5916.7	-647.63	3213.7	-357.75
5c	6270.2	-856.74	4661.6	-583.84	3792	-419.21	2477.2	-270.08
5d	22990	-2580.4	18570	-1873.2	17103	-1686.7	16482	-1577.8
5e	17672	-2002.2	12900	-1350.8	11404	-1149.7	7350.3	-721.74

Table 4.6: Falkenhagen Coefficient ("A" $dm^{3/2}$ $mol^{-1/2}$) and Jones-Dole Coefficient ("B" dm^3 mol^{-1}) for Dihydropyrimidin-2(1H)-ones at different Temperatures

Comp.	298.15 K		303.15 K		308.15 K		313.15 K	
	A	B	A	B	A	B	A	B
5a	0.4816	-2.3168	0.244	-0.5472	0.2107	-0.1839	0.1611	0.3123
5b	0.6582	-3.3455	0.4409	-1.9715	0.4233	-1.6953	0.3176	-0.7744
5c	0.5883	-2.917	0.3069	-0.9092	0.2781	-0.6216	0.2361	-0.2063
5d	1.5119	-9.9792	0.9521	-5.8451	0.6351	-3.7134	0.6052	-3.0667
5e	1.2748	-8.1627	0.6297	-3.3316	0.5755	-2.8596	0.4823	-2.0182

Figure 4.4: Graph of $\eta_r - 1/\sqrt{C}$ Vs $\sqrt{C}$ for Dihydropyrimidin-2(1H)-ones (5a-5e) at different Temperatures

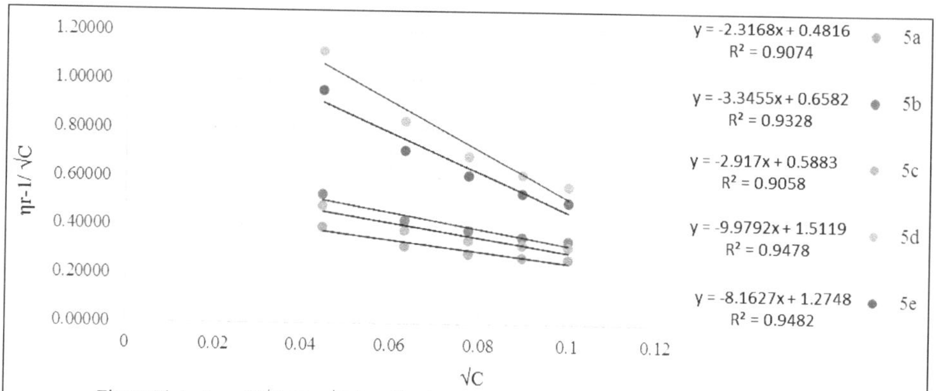

Figure:Plot of ηr-1/√C Vs √C for dihydropyrimidin-2(1H)-one 5a to 5e at 298.15 K

298.15 K

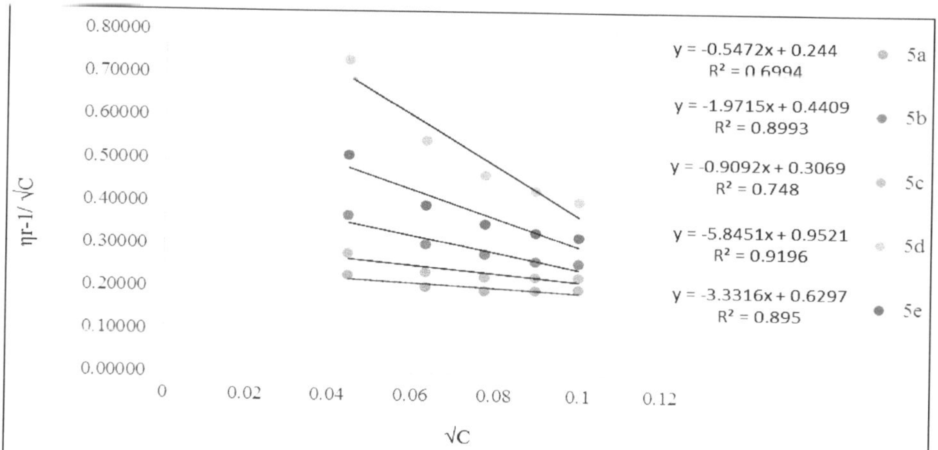

Figure:Plot of ηr-1/√C Vs √C dihydropyrimidin-2(1H)-one 5a to 5e at 303.15 K

303.15 K

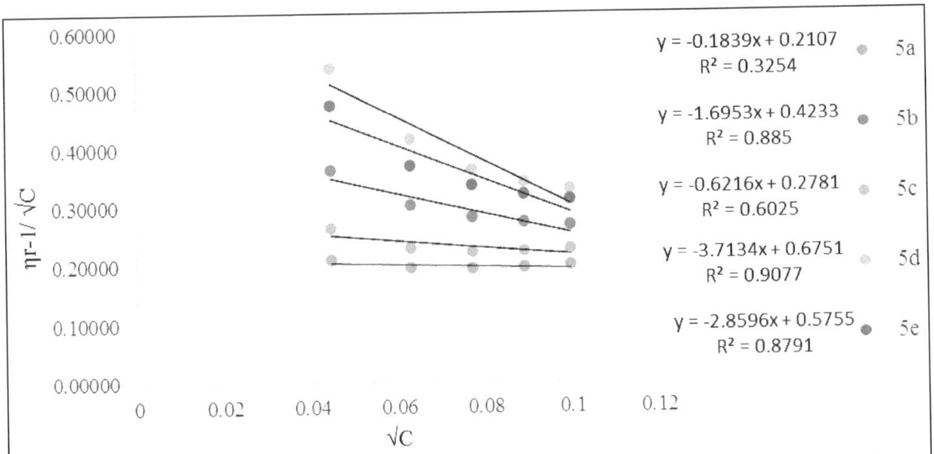

Figure:Plot of ηr-1/ √C Vs √C dihydropyrimidin-2(1H)-one 5a to 5e at 308.15 K

308.15 K

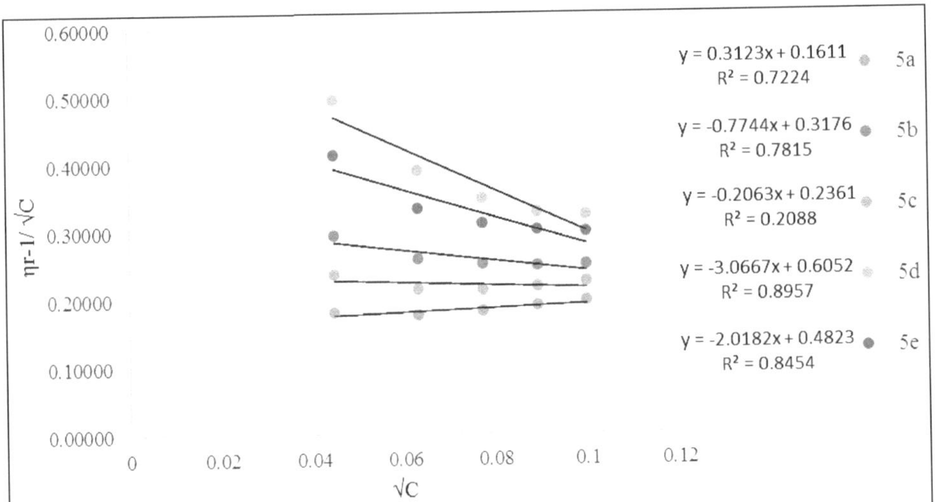

Figure:Plot of ηr-1/ √C Vs √C dihydropyrimidin-2(1H)-one 5a to 5e at 313.15 K

313.15 K

4.4.3 Density, Viscosity of Dihydropyrimidine Carboxylates (6a-6e)

Density, viscosity and relative viscosities of various substituted dihydropyrimidine carboxylates (6a-6e) are investigated from molar concentration 0.002 to 0.01 mol dm^{-3} in 60 per cent (v/v) aqueous dimethyl sulphoxide at 298.15, 303.15, 308.15 and 313.15 K temperature were systematically presented in Table 4.7. From "ρ" values, "$Ø_v$" values of dihydropyrimidine carboxylates have been estimated by using Masson equation as tabulated in Table 4.7. At 298.15, 303.15,

308.15, 313.15 K temperatures it was noticed that there is slight increase with concentration and reduces with elevation in temperature. This trend was proved from the values represented in Table 4.7 followed by study of apparent molar volume for all substituted pyrimidine carboxylates under investigation which showed increase with increase in concentration and temperature. The apparent molar volume was determined using equation as shown below

$$\varnothing_v = \varnothing_v^0 + S_v C^{1/2}$$

Table 4.7: Density (ρ), Apparent Molar Volume ($\varnothing_v$), Viscosity (η) and Relative Viscosity (η_r) of Pyrimidine Carboxylates (6a-6e) in 60 per cent Aqueous DMSO at different Temperatures

Comp	Conc (mol dm^{-3})	√C	Density "ρ" (g cm^{-3})	$\varnothing_v$ (cm^3.mol^{-1})	Viscosity "η" (cP)	Relative Viscosity "η_r"
			298.15 K			
6a	0.002	0.0447	1.08484	-844.455	3.60534	1.06965
	0.004	0.0632	1.08671	-726.6269	3.61711	1.07314
	0.006	0.0775	1.08698	-440.9983	3.62356	1.07505
	0.008	0.0894	1.08734	-308.5770	3.63032	1.07706
	0.01	0.1	1.08796	-253.1436	3.63795	1.07932
6b	0.002	0.0447	1.08514	-965.9379	3.66734	1.08804
	0.004	0.0632	1.08524	-370.0317	3.67322	1.08979
	0.006	0.0775	1.08551	-197.5713	3.67968	1.09170
	0.008	0.0894	1.0859	-125.1984	3.68656	1.09374
	0.01	0.1	1.08585	-41.1265	3.69193	1.09534
6c	0.002	0.0447	1.08493	-854.1554	3.62782	1.07632
	0.004	0.0632	1.08524	-355.2506	3.63440	1.07827
	0.006	0.0775	1.08551	-182.7901	3.64085	1.08018
	0.008	0.0894	1.0859	-110.4173	3.64771	1.08222
	0.01	0.1	1.08585	-26.3454	3.65309	1.08381
6d	0.002	0.0447	1.08628	-1468.035	3.69340	1.09577
	0.004	0.0632	1.0867	-682.7454	3.70038	1.09784
	0.006	0.0775	1.08713	-422.5219	3.70740	1.09993
	0.008	0.0894	1.0873	-262.3859	3.71353	1.10175
	0.01	0.1	1.0875	-169.0758	3.71977	1.10360
6e	0.002	0.0447	1.08605	-1349.7858	3.67596	1.09060
	0.004	0.0632	1.08612	-536.7816	3.68175	1.09232
	0.006	0.0775	1.08717	-416.6710	3.69087	1.09502
	0.008	0.0894	1.08639	-145.2915	3.69377	1.09588
	0.01	0.1	1.087	-110.8751	3.70140	1.09815

Comp	Conc (mol dm^{-3})	√C	Density "ρ" (g cm^{-3})	$Ø_v$ (cm^3.mol^{-1})	Viscosity "η" (cP)	Relative Viscosity "$η_r$"
303.15 K						
6a	0.002	0.0447	1.08314	-571.1367	3.05702	1.02203
	0.004	0.0632	1.08384	-319.7848	3.06453	1.02454
	0.006	0.0775	1.0839	-137.3589	3.07025	1.02645
	0.008	0.0894	1.0845	-108.5678	3.07749	1.02888
	0.01	0.1	1.08455	-40.4309	3.08318	1.03078
6b	0.002	0.0447	1.08395	-928.559	3.11474	1.04133
	0.004	0.0632	1.08456	-469.1347	3.12204	1.04377
	0.006	0.0775	1.08556	-376.1031	3.13047	1.04659
	0.008	0.0894	1.08563	-222.083	3.13622	1.04851
	0.01	0.1	1.0865	-203.6597	3.14429	1.05121
6c	0.002	0.0447	1.08345	-682.5712	3.08560	1.03159
	0.004	0.0632	1.0842	-371.1094	3.09328	1.03415
	0.006	0.0775	1.08456	-207.1789	3.09985	1.03635
	0.008	0.0894	1.08495	-128.6815	3.10652	1.03858
	0.01	0.1	1.08556	-101.9279	3.11381	1.04102
6d	0.002	0.0447	1.08464	-1223.0978	3.17219	1.06054
	0.004	0.0632	1.085	-546.3528	3.17879	1.06274
	0.006	0.0775	1.08435	-165.1019	3.18243	1.06396
	0.008	0.0894	1.0852	-147.8704	3.19048	1.06665
	0.01	0.1	1.086	-132.9077	3.19838	1.06929
6e	0.002	0.0447	1.08384	-841.1688	3.14214	1.05049
	0.004	0.0632	1.08396	-293.8913	3.14803	1.05246
	0.006	0.0775	1.0842	-129.9608	3.15427	1.05454
	0.008	0.0894	1.08482	-91.9220	3.16162	1.05700
	0.01	0.1	1.08539	-64.4749	3.16883	1.05941
308.15 K						
6a	0.002	0.0447	1.07435	-381.89582	2.60890	1.00979
	0.004	0.0632	1.075	-213.7816	2.61598	1.01253
	0.006	0.0775	1.07549	-132.8905	2.62268	1.01513
	0.008	0.0894	1.0755	-36.5259	2.62821	1.01727
	0.01	0.1	1.0759	-15.0545	2.63470	1.01978
6b	0.002	0.0447	1.07495	-644.2496	2.66539	1.03166
	0.004	0.0632	1.07535	-278.0885	2.67189	1.03417
	0.006	0.0775	1.07611	-211.9540	2.67928	1.03703
	0.008	0.0894	1.07656	-142.7722	2.68591	1.03960
	0.01	0.1	1.07686	-87.2834	2.69217	1.04202

Comp	Conc (mol dm^{-3})	$\sqrt{C}$	Density "ρ" (g cm^{-3})	$Ø_v$ (cm^3.mol^{-1})	Viscosity "η" (cP)	Relative Viscosity "η_r"
6c	0.002	0.0447	1.0747	-512.8397	2.63726	1.02077
	0.004	0.0632	1.07525	-239.8771	2.64412	1.02342
	0.006	0.0775	1.0761	-195.4889	2.65172	1.02636
	0.008	0.0894	1.0762	-85.9211	2.65747	1.02859
	0.01	0.1	1.07688	-74.2356	2.66466	1.03138
6d	0.002	0.0447	1.07579	-1010.9858	2.72254	1.05378
	0.004	0.0632	1.07585	-369.8891	2.72820	1.05597
	0.006	0.0775	1.07589	-153.0835	2.73380	1.05814
	0.008	0.0894	1.07635	-93.6100	2.74048	1.06072
	0.01	0.1	1.07676	-53.2659	2.74704	1.06326
6e	0.002	0.0447	1.0751	-677.3351	2.69328	1.04245
	0.004	0.0632	1.07525	-217.9755	2.69916	1.04473
	0.006	0.0775	1.0757	-111.4549	2.70579	1.04730
	0.008	0.0894	1.07635	-81.49427	2.71294	1.05006
	0.01	0.1	1.07694	-57.9259	2.71994	1.05277
313.15 K						
6a	0.002	0.0447	1.06642	-188.2114	2.26187	1.00820
	0.004	0.0632	1.06728	-166.3745	2.26917	1.01146
	0.006	0.0775	1.06775	-98.0894	2.27564	1.01434
	0.008	0.0894	1.0679	-26.4046	2.28143	1.01692
	0.01	0.1	1.0683	-6.8577	2.28776	1.01974
6b	0.002	0.0447	1.06685	-372.6376	2.28466	1.01836
	0.004	0.0632	1.06725	-141.9720	2.29099	1.02118
	0.006	0.0775	1.06795	112.0113	2.29796	1.02429
	0.008	0.0894	1.06871	-104.0701	2.30507	1.02746
	0.01	0.1	1.06931	-84.2885	2.31185	1.03048
6c	0.002	0.0447	1.06662	-249.6868	2.27323	1.01327
	0.004	0.0632	1.06752	-190.3077	2.28062	1.01656
	0.006	0.0775	1.06825	-143.9222	2.28766	1.01970
	0.008	0.0894	1.0686	-76.1481	2.29388	1.02247
	0.01	0.1	1.0688	-21.4053	2.29979	1.02510
6d	0.002	0.0447	1.06813	-948.4415	2.30930	1.02934
	0.004	0.0632	1.06825	-351.7392	2.31503	1.03190
	0.006	0.0775	1.06829	-140.3244	2.32059	1.03438
	0.008	0.0894	1.0684	-42.8294	2.32631	1.03692
	0.01	0.1	1.068935	-24.2209	2.33295	1.03988
6e	0.002	0.0447	1.06725	-523.2758	2.29646	1.02362

Comp	Conc (mol dm^{-3})	√C	Density "ρ" (g cm^{-3})	$Ø_v$ (cm^3.mol^{-1})	Viscosity "η" (cP)	Relative Viscosity "$η_r$"
	0.004	0.0632	1.06756	-177.6372	2.30259	1.02635
	0.006	0.0775	1.06795	-74.9384	2.30891	1.02917
	0.008	0.0894	1.06839	-29.4549	2.31534	1.03203
	0.01	0.1	1.06925	-41.5842	2.32268	1.03531

Figure 4.5: Graph of √C Vs $Ø_v$ for Dihydropyrimidine Carboxylates (6a-6e) at different Temperatures

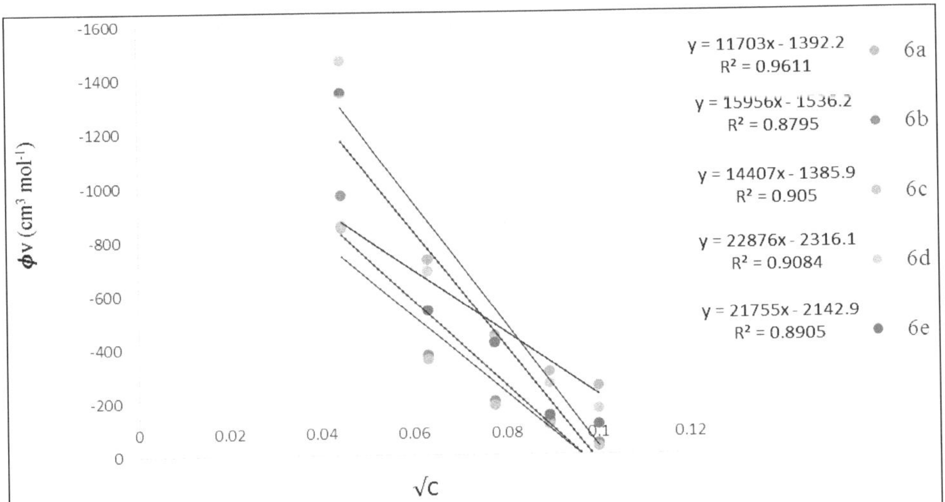

Figure: Plot of √C Vs ϕv dihydropyrimidine carboxylate 6a to 6e at 298.15 K

298.15 K

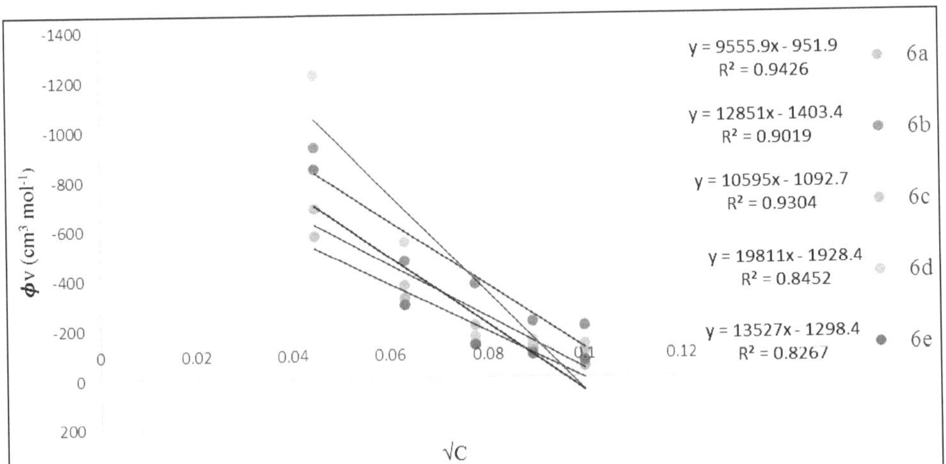

Figure: Plot of √C Vs ϕv for dihydropyrimidine carboxylate 6a to 6e at 303.15 K

303.15 K

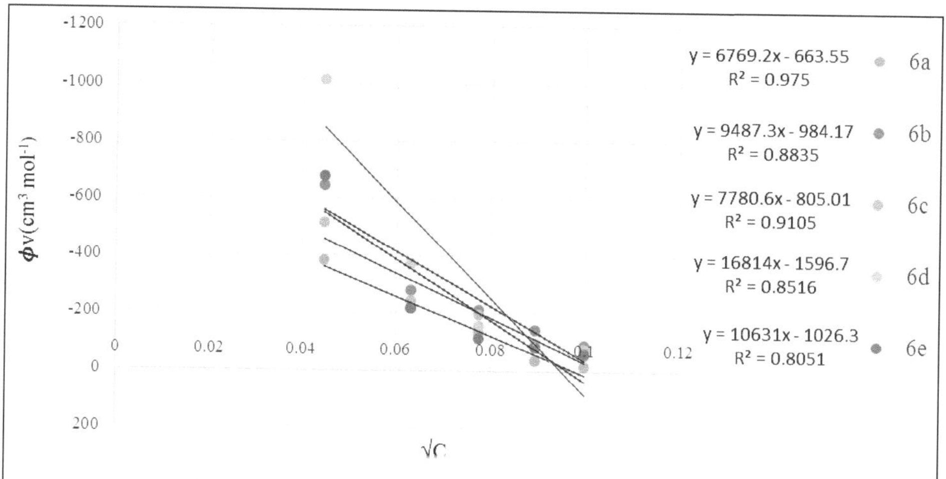

Figure: Plot of √C Vs ϕv for dihydropyrimidine carboxylate 6a to 6e at 308.15 K

308.15 K

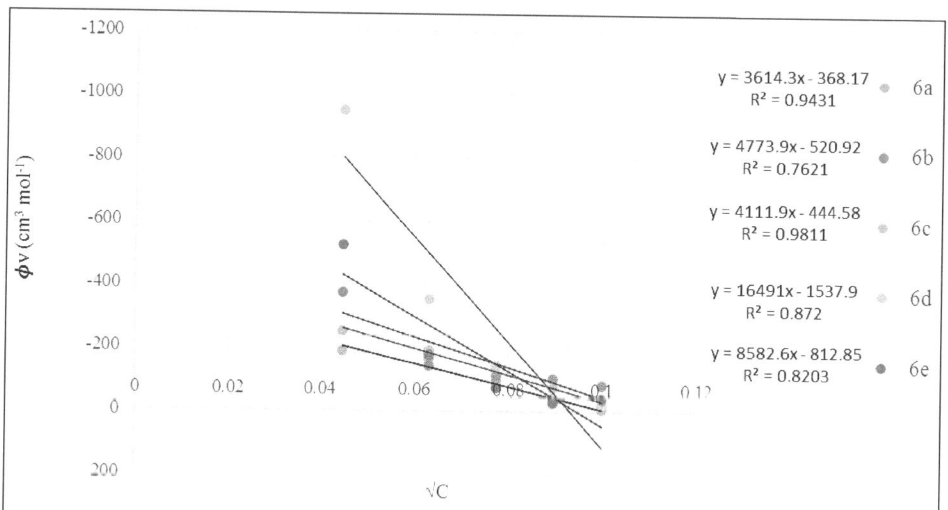

Figure: Plot of √C Vs ϕv for dihydropyrimidine carboxylate 6a to 6e at 313.15 K

313.15 K

The limiting apparent molar volume Ø_v^0 of dihydropyrimidine carboxylates at 298.15, 303.15, 308.15 and 313.15 K from intercept obtained by plotting extrapolating graph of apparent molar volume "Ø_v" against square root of concentration as shown in Figure 4.5 to infinite dilution and correlation coefficient ranging 0.7621-0.9811. The value of semi empirical parameter "S_v" is obtained from slope. The values of Ø_v^0 and S_v for compounds of dihydropyrimidine are arranged systematically in Table 4.8. The values of "S_v" are highly positive pointing out existence of strong solute-solute interactions. "Sv" values drops out with elevation of temperature explored that the solute-solute interaction were strong at stubby temperature 298.15 K and weakens with rise in temperature (313.15 K) [69].

Figure 4.6: Graph of $\eta_r - 1/\sqrt{C}$ Vs $\sqrt{C}$ for Pyrimidine Carboxylates (6a-6e) at different Temperatures

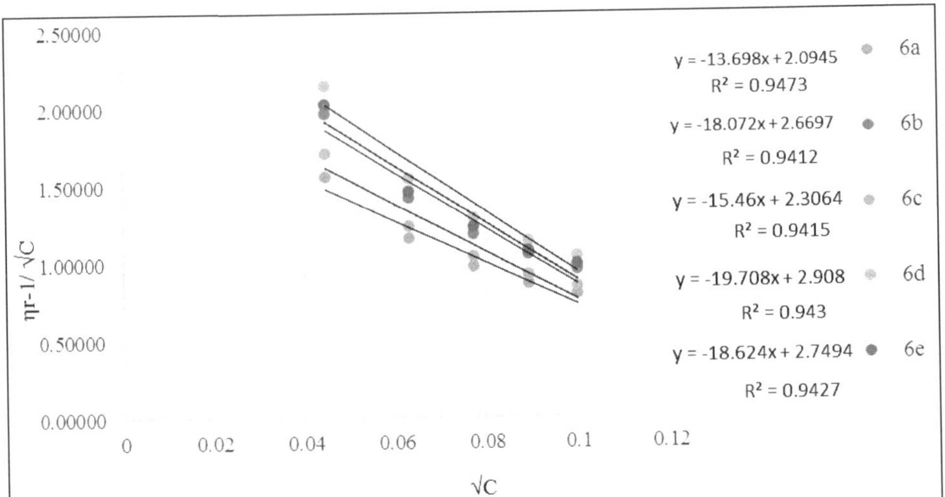

Figure:Plot of ηr-1/√C Vs √C for pyrimidine carboxylate 6a to 6e at 298 K

298.15 K

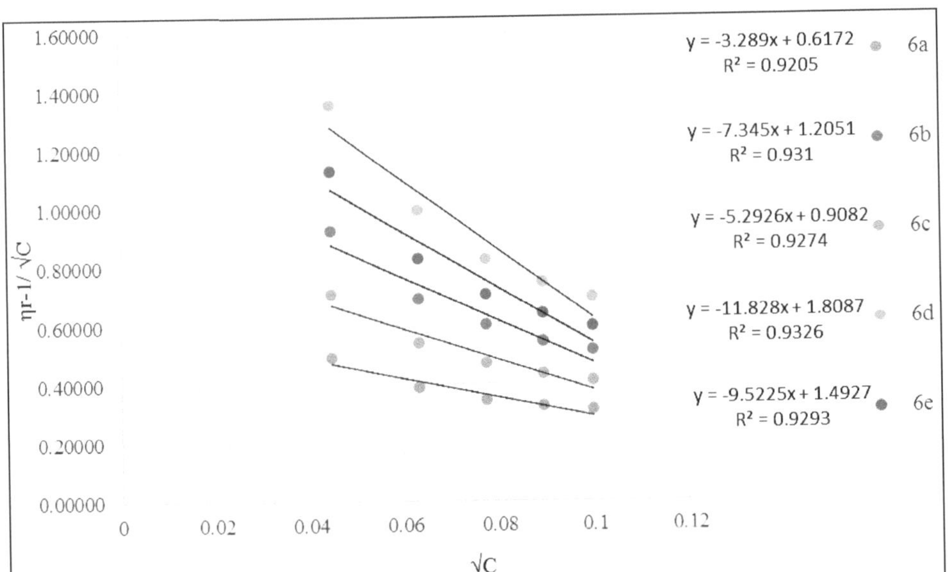

Figure:Plot of ηr-1/√C Vs √C for pyrimidine carboxylate 6a to 6e at 303.15 K

303.15 K

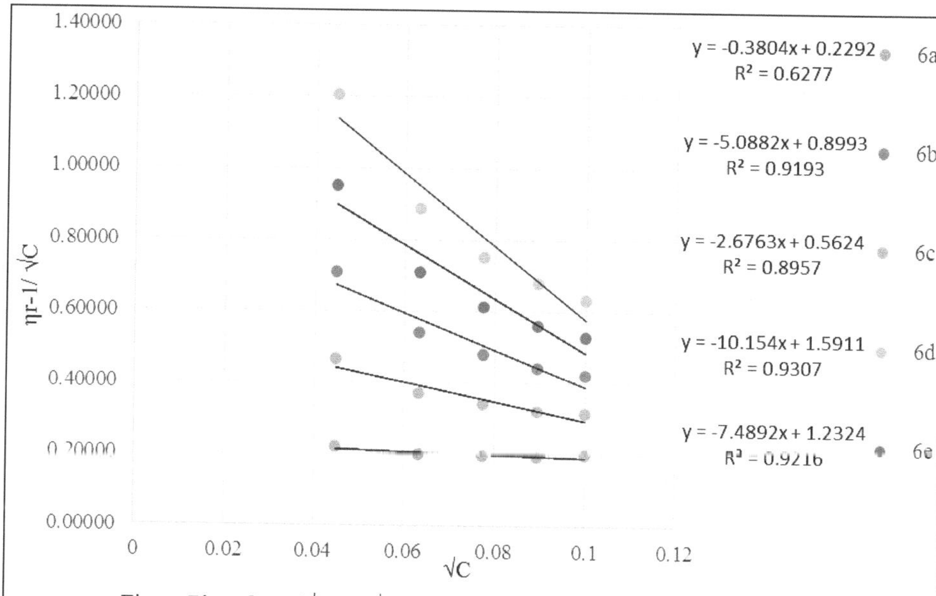

Figure:Plot of ηr-1/√C Vs √C pyrimidine carboxylate 6a to 6e at 308.15 K

308.15 K

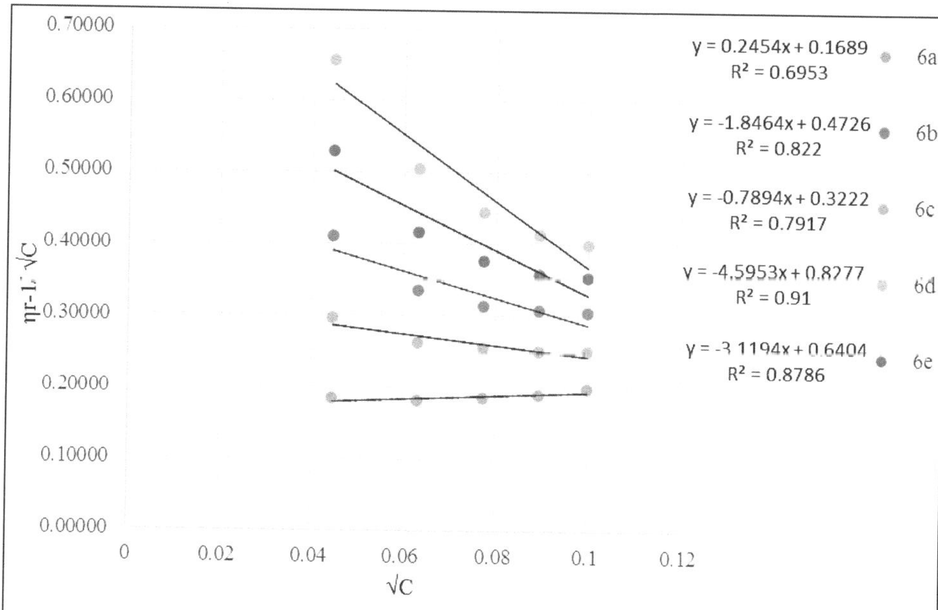

Figure:Plot of ηr-1/√C Vs √C for pyrimidine carboxylate 6a to 6e at 313.15 K

313.15 K

Viscosity data for derived compounds of dihydropyrimidine carboxylates were examined by using the Jones-Dole equation [24]. The values of viscometric parameters such as Falkenhagen coefficient "A" determined from intercept and Jones-Dole coefficient "B" from slope at various temperatures (298.15, 303.15, 308.15 and 313.15 K) from plotting graph of $\eta_r - 1/\sqrt{C}$ against $\sqrt{C}$ as shown in Figure 4.6 to infinite dilution and correlation coefficient ranging 0.6277-0.9473. "A" and "B" for dihydropyrimidine carboxylates and their compounds in Table 4.8. of "A" values are positive and represent occurrence of solute-solute interactions which ultimately decreases with increase in temperature which points that solute-solute interaction diminishes with increase in temperature.[70].

Table 4.8: Limiting Apparent Molar Volume "$Ø_v^0$"($cm^3\ mol^{-1}$) and Semi Empirical Parameter (Experimental slope "S_v") of Pyrimidine Carboxylates at different Temperatures

Comp.	298.15 K		303.15 K		308.15 K		313.15 K	
	S_v	$Ø_v^0$	S_v	$Ø_v^0$	S_v	$Ø_v^0$	S_v	$Ø_v^0$
6a	11703	-1392.2	9555.9	-951.9	6769.2	-663.55	3614.3	-368.17
6b	15956	-1536.2	12851	-1403.4	9487.3	-984.17	4773.9	-520.92
6c	14407	-1385.9	10595	-1092.7	7780.6	-805.01	4111.9	-444.58
6d	22876	-2316.1	19811	-1928.4	16814	-1596.7	16491	-1537.9
6e	21755	-2142.9	13527	-1298.4	10631	-1026.3	8582.6	-812.85

Table 4.9: Falkenhagen Coefficient ("A" $dm^{3/2}\ mol^{-1}$) and Jones-Dole Coefficient ("B" $dm^3\ mol^{-1}$) for Pyrimidine Carboxylates at different Temperatures

Comp.	298.15 K		303.15 K		308.15 K		313.15 K	
	A	B	A	B	A	B	A	B
6a	2.0945	-13.698	0.6172	-3.289	0.2292	-0.3804	0.1689	0.2454
6b	2.6697	-18.072	1.2051	-7.345	0.8993	-5.0882	0.4726	-1.8464
6c	2.3064	-15.46	0.9082	-5.2926	0.5624	-2.6763	0.3222	-0.7894
6d	2.9080	-19.708	1.8087	-11.828	1.5911	-10.154	0.8277	-4.5953
6e	2.7494	-18.624	1.4927	-9.5225	1.2324	-7.4892	0.6404	-3.1194

Hence solute-solute interaction among the five members between dihydropyrimidine carboxylates follow the trend having decreasing order : 6d > 6e > 6b > 6c > 6a.

4.5 Conclusion

Different parameters such as density, viscosity, apparent molar volume, limiting apparent molar volume, experimental slope, Falkenhagen coefficient ("A") and Jones-Dole coefficient ("B") for dihydropyrimidine carbonitriles (4a-4e), dihydropyrimidin-2(1H)-ones (5a-5e) and dihydropyrimidine carboxylates (6a-6e) and their related derivatives in 60 per cent (v/v) aqueous dimethyl sulphoxide for

0.002, 0.004, 0.006, 0.008 and 0.01 M concentration were thoroughly studied at 298.15, 303.15, 308.15 and 313.15 K temperatures.

From the investigation inputs, we concluded in each of the cases that densities of all related derivatives in 60 per cent (v/v) aqueous DMSO increase with increase in concentration at 298.15, 303.15, 308.15, 313.15 K and have a decreasing trend with rise in temperature. The trend for increase in apparant molar volume with increase in concentration and temperature was proved from the data tabularized.

Large values of experimental slope "S_v" obtained gave an idea about existence of solute-solute interactions and formerly found that value of "S_v" decreases with increase in temperature acknowledge that solute-solute interaction becomes weaker and weaker.

Also resulted values of viscosity coefficient or Falkenhagen coefficient positive "A" decreases with elevation in temperature indicating clearly that there is existence of solute-solute interactions and diminishes or weakens with rise in temperature. The Jones-Dole coefficient "B" is negative in magnitude and elevates with inflation of temperature clearly specify the existence of additional solute-solvent interactions at temperatures by elevation.

Also by contemporary study we resulted strong solute-solute interactions dihydropyrimidine carbonitriles than dihydropyrimidine carboxylates and to smaller extent in dihydropyrimidin-2(1H)-ones. Among dihydropyrimidine carbonitriles the decreasing trend in solute-solute interactions was found to follow the order: 4d > 4e > 4b > 4c > 4a, while in case of dihydropyrimidin-2(1H)-ones the decreasing trend follow the order of solute-solute interaction as 5d > 5e > 5b > 5c > 5a and finally in case of dihydropyrimidine carboxylates, the decreasing order is found to be 6d > 6e > 6b > 6c > 6a. This trend may be characterized due to high molecular association in dihydropyrimidine carbonitriles than dihydropyrimidine carboxylates and lastly in dihydropyrimidin-2(1H)-ones.

4.6 References

1. R. R. Nolasco, L.A. Galicia-Luna, O. Elizalde Solis, *J. Chem. Thermodyn.* 44, **2012**, 133-147.
2. M. T. Zafarani-Moattar, F. Izadi, *J. Chem. Thermodyn.* 43, **2011**, 552-561.
3. A. Chandra, B. Bagchi, *J. Phys. Chem.* B 104, **2000**, 9067-9080.
4. A. Chandra, B. Bagchi, *J. Chem. Phys.*113, **2000**, 3226-3232.
5. P. K. Banipal, T. S. Banipal, B. S. Lark and J. C. Ahluwalia, *J. Chem. Soc. Faraday Trans.*, 93, **1997**, 81.
6. S. Das, S. K. Ray and D. K. Hzara, *Indian J. Chem.*, 41, **2002**, 1812.
7. M. L. Parmar, R. K. Awasthi and M. K. Guleria, *J. Chem. Sci.*, 116, **2004**, 33-38.
8. M. L. Parmar and D. S. Banyal, *Indian J. Chem.*, 44A, **2005**, 1582.
9. M. L. Parmar and M. K. Guleria, *J. Indian Chem. Soc.*, 82, **2005**, 648.

10. M. Singh, *J. Chem. Sci.,* 118, **2006**, 269.
11. D. M. Swenson and Wooley, *J. Chem. Thermodyn.*, 40, **2008,** 54.
12. C. Marignac, *Ann. Chim.* (Paris), 22, **1871**, 415.
13. M. L. Parmar and M.K. Guleria, *Indian J. Chem.*, 48A, **2009**, 806-811.
14. A. B. Nikumbh, G. K. Kulkarni, M. Rathi and R. C. Bhujbal, *J. Harm. Res.* 4(1), **2015**, 27-38.
15. J. L. Shen, Zhi-Fen Li, Bao-Huai Wang and You-Min Zhang, *J. Chem. Thermodyn.* 32, **2000**, 805-819.
16. M. J. Iqbal and M. A. Chaudhry, *J. Chem. Thermodyn.* 42, **2010**, 951-956.
17. S. S. Dhondge, S. P. Zodape and D. Parwate, *J. Chem. Thermodyn.* 48, **2012**, 207-212.
18. S. S. Dhondge, J. Paliwal, *J. Chem. Thermodyn.* 58, **2013**, 149-157.
19. D. O. Masson, *Phil. Mag.,* 8, **1929**, 218.
20. a) C. Redlich and D.M. Meyer, *Chem. Rev.* 64, **1964**, 221. b) O. Redlich, *J. Phys. Chem.*, 67, **1968**, 496.
21. B. B. Owen and S. R. Brinkley, Ann. N. Y. *Acad. Sci.*, 51, **1949**, 753.
22. A. F. Scott, *J. Phys. Chem.*, 35, **1931**, 2315 and 3379.
23. W. Geffeken, *J. Phys. Chem.* Abt. A, 155, **1931**, 1.
24. G. Jones and M. Dole, *J. Am. Chem. Soc.*, 51 (10), **1929**, 2950-2964.
25. J. Wu, Z. Xu, Z. Liu and B. Wang, *J. Chem. Eng. Data*, 50(3), **2005**, 966–968.
26. R. M. Diguilio, R. J. Lee, S. T. Schaeffer, L. L. Brasher and A. S. Teja *J. Chem. Eng. Data*, 37(2), **1992**, 239-242.
27. Y. Su, J. Yao, C. Wang and H. Li, *J. Chem. Eng. Data*, 57 (2), **2012**, 298–308.
28. I. S. Khattab, F. Bandarkar, M. A. A. Fakhree and A. Jouyban, *Korean J. Chem. Eng.*, 29(6), **2012**, 812-817.
29. P. J. Carvalho, C. F. G. Fonseca, M. C. J. Moita, A. F. S. Santos and J.A. P. Courtino, *J. Chem. Eng. Data*, 60 (12), **2015**, 3721–3737.
30. B. Gonzalez, N. Calvar, E. Gomez, A. Dominguez, *J. Chem. Therm.,* 39(12), **2007**, 1578-1588.
31. M. H. Rausch, L. Kretschmer, S. Will, A. Leipertz and A. P. Froba, *J. Chem. Eng. Data*, 60 (12), **2015**, 3759–3765.
32. E. Vogel, *J. Sol. Chem.,* 41(8), **2012**, 1422-1445.
33. A. Hartano and H. F. Svendsen, *J. Chem. Therm.,* 41(9), **2009**, 973-979.
34. U. Domanska and M. Krolikowska, *J. Sol. Chem,* 41(8), **2012**, 1422-1445.
35. R. J. Correia, J. Kestin, and H. E. Khalifa, *J. Chem. Eng. Data*, 25 (3), **1980**, 201–206.

36. M. Singh , H. Chand and K. C. Gupta, *Chem. Biodivers.*, 2(6), 2005, 809-824.
37. Y. P. Hsieh, R. B. Leron, A. N. Soriano, Alvin. R. Caparanga, M. Li, *J. Chem. Eng. Japan.*, 45(12), **2014**, 939-947.
38. S. Han, C. Du, X. Jian, L. Meng, X. R. Jie, W. Jian, and H. Zhao, *J. Chem. Eng. Data*, 61 (2), **2016**, 731–739.
39. T. A. Graber, H. R. Galleguillos, C. Céspedes and M. E. Taboada, *J. Chem. Eng. Data*, 49 (5), **2004**, 1254-1257.
40. A. Bhattacharjee, C. Varanda, M. G. Freire, S. Matted, L. Santos, I. M. Marracho and J. A. P. Coutinho, *J. Chem. Eng. Data*, 57 (12), **2012**, 3473–3482.
41. N. V. Sastry and I. R. Ravalji, *J. Chem. Eng. Data*, 61(11), **2016**, 3834-3448.
42. S. K. Mylona and M. J. Assael, *J. Phy. and Chem. Ref. Data.*, 42(3), **2013**, 1-6.
43. S. M. Nayeem, M. Kondaiah, K. Sreekanth, and D. K. Rao, *J. Therm.*, 2014, **2014**, 1-13.
44. D. I. Sagdeev, M. G. Fomina, G. K. Mukhamedzyanov and I. M. Abdulagatov, *Thermochemica Acta*, 592, **2014**, 73-85.
45. J. Jacquemin, P. Husson, A. A. H. Paduaand and V. Majer, *Green Chem.*, 8(2), **2006**, 172-180.
46. K. Rajagopal, G. R, Renold, *Inter. J. PharmTech Res.*, 8 (8), **2015**, 180-195.
47. R. V. Entrolizo, *Sci. Coop. Work. Eng. Bran.*, **2015**, 163-167.
48. K. Saravanakumar and T. R. Kunbendran, *Res. J. Chem. Sci.*, 2 (4), **2012**, 50-56.
49. C. Ujjwala, B. Berad, P. Agrawal, B. Meshram, *J. Glo. Sci.*, 4(2), **2015**, 1530-1545.
50. A. A. Sheikha, P. A. Ali Khana, U. K. Asemaa and M. Asif, *J. Med. Chem. drug. dis.*, 3(2), **2017,** 223-227.
51. G. A. Dopazo1, M. G. Temes, D. L. Lopez and J. C. Mejuto1, *Med. J. Chem,*. 3(4), **2014**, 972-986.
52. R. E. Pavai and S. Renuka, *Int.J. Res. Pur. Chem.*, 1(2), **2011**, 6-10.
53. K. A. Kumar and Dr. C. Srinivasu, *Int. J. Inno. Res. Sci. Eng. Tech.*, 3(7), **2014**, 14880-14889.
54. S. Santhi and J. Madhumitha, *Int. J. Adv. Chem.*, 2(1), **2014**, 12-16.
55. U. K. Asima, M. ASif P. D. Netankar and A. A. Shaikh, Int. *J. Res. Chem. Env.*, 7(1), **2017**, 14-16.
56. E. Kartzmark, *Can. J. Chem.*, 50, **1972**, 2845-2850.
57. P. W. Derks, K, J. Hogendoorn and V. Geert, *J. Chem. Eng. Data.*, 51, **2006**, 777.
58. P. W. Derks, E. S. Hamborg, K, J. Hogendoorn, J. P. Niederer and G. F. Versteeg, *J. Chem. Eng. Data.*, 53, **2008**, 1179-1185.
59. G. S. Reddy, and M. M. Reddy, *Int. J. Pharm. Bio. Sci.*, 5(1), **2014**, 1064-1073.

60. V. Gupta. U. Magotra, A. Sharma and M. Sharma, *J. Chem. Env. Sci. App.*, 1(2), **2015**, 111-119.
61. S. S. Kulkarni and U V. Khadke, *Ind. J. Mat. Sci.*, 2016, **2016**, 1-6.
62. E. D. Dikio, S. M. Nilana, D. A. Isabirye and E. E. Ebanso, *Int. J Electrochem. Sci*, 7, **2012**, 1110-11122.
63. M. T. Andreuccetti, B. S. Litte and J. V. H. Angelo, *O. Papel,*. 72, **2011**, 52-57.
64. G. Bolat G. Lisa and I. M. Popa, *Sci. Stud. Res.*, 6(2), **2005**, 181-190.
65. G. S. Sanap, S. B. Ingole, S. G. Shankarwar, and A. G. Shankarwar., *IOSRJ. Eng.*, 6(9), **2016**, 25-29.
66. Z. P. Visak, A. G. M. Feirrara, and I. M. A. Fonsica, *J. Chem. Eng. Data.*, 45, **2002**, 926-931.
67. G. S. Reddy, and M. M. Reddy, *Int. J. Pharm. Bio. Sci.*, 5(11), **2013**, 644-648.
68. M. V. Rathnam, K. R. Bhanushali, R. T. Sayed and M. S. S. Kumar., *Eur. Chem. Bull.*, 2(7), **2013**, 434-444.
69. R. B. Dhake, *Glo. J. Res. Ana.*,5(2), **2016**, 178-179.
70. J. P. Singh, and R. Sharma, *Int. J. Eng. Res. Dev.*, 5(11), **2013**, 48-51.
71. K.Das and M. N. Roy, *Ind. J. Adv. Chem. Sci.*, 4(2), **2016**, 138-148.
72. M. Asif, S. S. Kasim, S. Parvin and A. A. Sheikh, *Int. J. Chem. Eng. Res.*, 8(1), **2016**, 47-50.
73. R. R. Naik, S. S. Bhawankar and S. N. Kachane, *J. Sci. Tech.*2(3), **2015**, 5-7.
74. P. G. Tello, F. Kamacho and G. Blazquez, *J. Chem. Eng. Data.*, 39(3),**1994**, *611-614.*
75. G. P. Borse, U. G. Deshpande, S. R. Patil and A. R. Hirey, *Der. Chemica. Senica.*, 3(6), **2012**, 1438-1443.
76. K. C. Patil and C. M. Dudhe, *Der Pharma. Chemica.*, 7(12), **2015**, 219-226.
77. V. D. Bhandakkar. *Arch. App. Sci. Res.*,8(3), **2016**, 16-21.
78. E. Obermeier, S. Fischer and D. Bohne, *Ber. Buns. Phy. Chem.*, 89, **1985**, 805-809.
79. K. Kaur and K. C. Juglan, *J. Chem. Pharm. Res.*, 8(7), **2016**, 49-53.
80. C. S. Selvi, S. Ravichandran, K. Rathin and C. Thenmozhi, *Proc. Indian Natn. Sci. Acad., 79(3)*, **2013**, 445-449.
82. J. S. Aher, M. R. Gaware, A. B. Sawant, S. V. Patil and D. D. Lokhande

Res. world., 4(4), **2016,** 120-124.

83. D. D. Lokhande, J. S. Aher, M. R. Gaware and A. V. Kardel, *Scho. Res. J. Interdis. stud.*, 6(21), **2017,** 92-97.
84. J. S. Aher, M. R. Gaware, D. D Lokhande and B. N. Shelke, *Scho. Res. J. Interdis. stud.*, 6(21), **2017,** 137-141.

Index

9 789359 191584

Printed at Repro India Ltd.